Global Sustainability

The world is in the midst of a transition to a form of global society. Factors such as economic globalization, communications technology, human mobility, cultural influence, and environmental change pull towards a unitary system, even as forces act against this connectivity. We live in a unique historical moment at which human activity can irreversibly alter the planet and its capacity to support life.

Unprecedented levels of wealth, technology, and institutional capacity can forge a just, peaceful, and ecologically resilient future. However, the authors argue, social polarization, geo-political conflict, and environmental degradation are threatening the long-term well-being of humanity and the planet. *Global Sustainability* explores the alternative futures that could emerge from the resolution of these antagonisms.

Based on extensive international and interdisciplinary research, the book analyzes the perils of market-driven scenarios and considers the possibility of the failure of conventional approaches. It identifies goals and strategies for "bending the curve" of development toward an environmentally and socially sustainable global future. It will be essential reading for all scholars and professionals interested in the future of the environment, international affairs, and sustainable development.

Gilberto C. Gallopín has been Director of the Systems for Sustainable Development Programme of the SEI based in Stockholm, Sweden; Leader of the Land Use Program at CIAT based in Cali, Columbia; Senior Fellow of the IISD, Winnipeg, Canada; Senior Expert on Environment and Development in the IIASA, Austria; Full Professor at the University of Buenos Aires and at the Fundacion Bariloche Argentina, as well as the Executive President of the latter. He is a founder member of the "Dialogue for a New World." Until early 1991 he was the Director of the Ecological Systems Analysis Group (Argentina).

With a background in theoretical physics, **Paul D. Raskin** founded the Tellus Institute in 1976, where as President he directs an organization of 50 professionals working on environment, resources, and development issues throughout the world. In 1989 he formed the Boston Centre of the Stockholm Environment Institute, which is hosted by Tellus and works in concert with the SEI. His current research focuses on the requirements for transition to sustainability at global, regional, national, and local levels. In 1995 he organized the Global Scenario Group, an international and interdisciplinary body whose major studies – *Branch Points, Bending the Curve*, and *Great Transitions* – are the inspiration for this book.

Routledge/SEI global environment and development series
Edited by Arno Rosemarin

Global Sustainability

Bending the curve

Gilberto C. Gallopín and Paul D. Raskin

London and New York

First published 2002
by Routledge
11 New Fetter Lane, London EC4P 4EE

Simultaneously published in the USA and Canada
by Routledge
29 West 35th Street, New York, NY 10001

Routledge is an imprint of the Taylor & Francis Group

© 2002 Stockholm Environment Institute

Typeset in 10/11.5 Garamond by
Newgen Imaging Systems (P) Ltd, Chennai, India
Printed and bound in Great Britain by
The Cromwell Press, Trowbridge, Wiltshire

British Library Cataloguing in Publication Data
A catalogue record for this book is available
from the British Library

Library of Congress Cataloging in Publication Data
A catalog record for this book has been requested

ISBN 0-415-26592-4

Contents

Figures

Tables

Acknowledgments

Two names are on the cover, but the insights of many are inside. The *Global Scenario Group* (GSG) was its source and wellspring. Since 1995, the GSG has engaged a diverse international group in an examination of alternative global futures – and the lessons they hold for policy, values, and action.

We are grateful to each of our GSG colleagues present and past – Tariq Banuri, Michael Chadwick, Khaled Mohammed Fahmy, Tibor Farago, Nadezhda Gaponenko, Pablo Gutman, Gordon Goodman, Allen Hammond, Lailai Li, Roger Kasperson, Robert Kates, Sam Moyo, Madiodio Niasse, H.W.O. Okoth-Ogendo, Atiq Rahman, Setijati Sastrapradja, Katsuo Seiki, Nicholas Sonntag, Rob Swart, and Veerle Vandeweerd. We are particularly indebted to those who joined us as coauthors on the studies that were the foundation for this book: *The Sustainability Transition: Beyond Conventional Development* (Raskin *et al.* 1996), *Branch Points: Global Scenarios and Human Choice* (Gallopín *et al.* 1997) and *Bending the Curve: Toward Global Sustainability* (Raskin *et al.* 1998). We thank the Stockholm Environment Institute, Rockefeller Foundation, the Nippon Foundation and the United Nations Environment Programme for major funding for GSG activities over the years.

We are deeply indebted to our colleague Eric Kemp-Benedict for his outstanding contribution and hard work at every phase of the scientific analysis and document preparation. We also appreciate the contributions of Charles Heaps, Jack Sieber, Robert Margolis, and Gil Pontius. Faye Camardo brought her eagle eye to editing a challenging manuscript.

We benefited from all of these able colleagues and partners, but the authors alone take responsibility for any remaining oversights and lapses in judgment.

Chapter I

The challenge

Ours is an age of profound transformation and great uncertainty about the future. Over the last few centuries – a mere heartbeat of historic time – the human impact on the global environment has grown from diminutive to elephantine. A tightening web of environmental, economic, and social interactions progressively binds nations, regions, and communities into a single Earth system. The dynamics of the global whole increasingly govern the behavior of its parts.

This book aims to illuminate the character of the current global system, the dynamics driving it forward, and a spectrum of possible future states and pathways. But more, the endeavor is animated by a conviction. It assumes that informed human choice, mediated through governmental policies, civil initiatives and individual decisions, can shape the future in essential ways. While reflecting realistically on the perils for the future, we retain the optimism that there are attractive possibilities, as well, for humanity and the environment in the twenty-first century.

The human predicament

For most of history, the challenge for the human race has been survival against natural forces that often have been harsh and unpredictable. The power to shape, control, and transform nature evolved gradually over several million years. Then, the emergence of the capitalist economic system and the modern worldview accelerated the process of change. The pace of innovation in technology, growth in population, and transformation of the environment and the social order reached a crescendo in the two centuries since the industrial revolution. Since 1950 alone, global population has doubled, energy production has more than tripled and economic output has increased by a factor of nearly 7. The interdependent global system we observe today is a way-station in this sweeping process of growth, transformation, and expansion.

Inevitably, such a rapid growth must butt against the limits of a finite planet. A new and ominous feature of the current phase of history is that human impacts on the environment have reached global scales. Human claims on environmental resources and disruption of environment support systems exceed natural rates for the renewal of resources and capacities for absorbing a complex brew of wastes generated by human activity. The contradiction between the growth imperative of the modern world system and the constraints of a finite planet will be resolved. The critical question is, how?

With the end of the Cold War, the threat to civilization of massive use of weapons of mass destruction may have receded. But a new and subtler challenge that holds both exciting promises and troubling perils faces humanity in the twenty-first century – the challenge of creating a sustainable global civilization. The human species has the capacity to apply its technological wizardry and its foresight to fashion a transition to an environmentally sustainable and just global society. The possibilities are unprecedented for technological and economic progress to eradicate hunger, improve the human condition, enrich the human stock of knowledge and cultural achievement, and increase opportunity and choice.

Yet, a zeitgeist of apprehension about the kind of world that this generation will bequeath to its descendants is displacing the Enlightenment faith in progress and hope for the future. For the risks are ominous. High growth in population in poor regions and consumption in rich regions are increasing the size of the human footprint on nature. The global climate system is destabilizing, ecosystems are degrading, and the Earth's biological wealth is diminishing. Billions of the yet unborn may be consigned to an existence of poverty, hunger, and hardship. The destitution of multitudes amidst unprecedented levels of wealth for the privileged portends social unrest and violence with a global reach. Globally connected terrorist networks, exploiting and feeding the despair and anger of the dispossessed, challenge the very notion of a global civilization.

How humanity will cope with such challenges is not certain. Nor is the outcome determined – it will be influenced by individual and collective choices that we make. While it is widely perceived that current trends are ecologically and socially unsustainable, an alternative vision has yet to be well articulated and the suite of actions that could provide safe passage to a sustainable future has not been defined. Will we be able to pass on to our grandchildren a global society – and a planet – that is richer in possibilities than our present one, or will we leave a more impoverished Earth as a patrimony for future generations? Will human existence and human institutions such as families and communities be more secure or more fragile in the global society of the mid-twenty-first century?

Conventional development wisdom generally assumes the expansion of resource-intensive consumption and production patterns in industrialized countries, and their gradual extension to developing countries. A common theme is that societies everywhere will gradually converge toward common institutional and cultural assumptions in the context of globalizing economies (OECD 1997). This paradigm animates the programs of international banks, the discussions of world trade negotiations and the ideologies of prestigious thinkers and leaders. Yet, a critical pragmatic question is whether the extrapolation of market-driven globalization envisioned by the conventional development paradigm is feasible. A critical normative one is whether such a vision for global development is desirable.

To gauge the scale of increased environmental pressure, if the world's projected population of 9 billion people in the middle of the twenty-first century were to consume resources at the same level per person as in the United States today, world requirements would grow by very roughly a factor of 10. Can such resource-intensive lifestyles be maintained and extended to a growing population? Can conventional socioeconomic goals and environmental sustainability be simultaneously satisfied? Or would conventional development risk unacceptable deterioration of the resources and ecosystems of the biosphere, and social and economic instability?

War, social opposition, and stubborn traditionalism impeded the forward march of the ascendant market system in the past. But with the collapse of the socialist experiments in Russia and elsewhere, the expansion of global markets, and the heady advance of new technology, the millennial glee of cheerleaders for global capitalism who were anticipating a cornucopia for all in the new century was, perhaps, understandable (Schwartz and Leyden 1997). But such a sanguine prognosis is simplistic. Substantial hurdles must be overcome, wise policies fashioned and fundamental questions addressed. How will a growing human enterprise, one that already is significantly perturbing natural planetary processes, be reconciled with environmental limits? How will the deep social fissures between the North and the South, the rich and the poor, parochialism and globalism, be ameliorated? And the question posed by Socrates long ago remains: how shall we live?

Roots in the industrial revolution

The global system that unfolds before us in its multifarious dimensions is the culmination of an expansionist and transformative European capitalism that emerged over the last millennium. By liberating nascent human potential for innovation and ingenuity, capacity for greed and acquisitiveness, and hunger for liberty and modernism, the new system set in motion a perpetual revolution in values, institutions, technology, and knowledge.

This process was further accelerated by the industrial explosion, which continues to play out in the tumultuous technological, institutional, and cultural changes of our time. The prodigious growth in material consumption, human numbers, industrial production, and claims on land and the whole range of natural resources, is the culmination of the advance of industrial society toward a world system. The enclosure of lands, which displaced traditional livelihoods and brought common resource areas into the market nexus, the colonial period and the current market expansion are all manifestations of this growth imperative. Dynamic capitalism has transformed the societies at its center, while progressively incorporating those on the periphery – or marginalizing them.

A number of factors combined to form a powerful, growth-oriented, modernizing, and dominant world system, albeit with sad counterpoints in social disruption, loss of community and environmental degradation. The industrial revolution was catalyzed by an interlinked series of technological innovations that vastly increased labor productivity by substituting machines and inanimate energy for human craft and muscle power, and sharply improved the capacity to exploit and manipulate raw materials (Landes 1970). Technological and social change drove one another in a mutually conditioning process of system transformation. The value of possessive individualism became a secular religion sweeping away more traditional and community-oriented norms. In economic theory, but to a much lesser extent in practice, the modern individual was of a new species; a rational, informed and acquisitive agent in the free market. Material wants and needs were met, expanded and transformed in a continuing spiral of production and consumption. The principle of economic efficiency as the rational basis for a rational economy was associated with the private control of investment surplus, the free market and unfettered trade.

At the same time, a number of modern institutions, building on historic antecedents, gradually developed to regularize and reform the maturing capitalist system. A modern legal and constitutional framework arose to regulate economic conduct, guarantee contracts, and protect, to some extent, social and civil liberties. Meanwhile, oppositional institutions – labor unions, suffrage movements, and minority rights organizations – struggled for, and often won, better working conditions, democratic enfranchisement and social and economic opportunity for marginalized groups.

A parallel and reinforcing constellation of attitudes arose in religion (the Reformation), political philosophy, and modern science. The traditionalism of received dogma and birthright gave way to a modernism, which embraced values compatible with the industriousness and entrepreneurship of the new era. With roots in Judaeo–Christian attitudes toward nature, industrial society saw nature as a cornucopia for human domination, an essentially limitless wellspring of resources, space, and services. Spawned by the new order, the scientific revolution, in turn, greatly hastened the process of transformation, both by spinning off an endless stream of new technology and by altering human awareness of its place in nature and the cosmos.

The sustainability transition

The inexorable expansion of human activity unleashed by the industrial era was destined to reach a planetary phase. At the dawn of the new century, that era is upon us. Its most vivid expression is perhaps global environmental change. Where the critical environmental issues of 30 years ago, such as air pollution, were local, straightforward, and short term, the environmental agenda today includes issues – climate change is the epitome – that are global, complex, and long term. But global environmental transformation is one aspect of a unitary process of globalization, which also has economic, technological, cultural, and geo-political dimensions.

The concern for the long-term well-being of the planet and future generations has been captured in the notion of *sustainable development*. Sustainable development is an imprecise concept. Indeed, for some the very notion of development that is sustainable seems oxymoronic. Nevertheless, while precise technical definitions may be elusive, agreement on what is *not* sustainable is widespread. Like many powerful concepts, its ambiguity is constructive, allowing for wide discussion and debate on the content of certain broad principles – a commitment to reconciling environmental and social goals, and a concern for the rights of future generations. The sustainability paradigm increasingly infuses policy discussions, intergovernmental initiatives and even business philosophy. Moreover, by underscoring the importance of integrated and systemic perspectives and multi-generational time horizons, it gradually is influencing the scientific research agenda.

Sustainability is concerned with reconciling the long-term development of human society with the finite limits of the planet. Implicit in the notion of sustainability are such questions as: How shall we use the Earth? What kind of human society shall we build on it? How can we leave future generations a world with more opportunities rather than fewer? The classic formulation that sustainable development "meets the needs of the present without compromising the ability of future generations to meet their own needs" reflects these broad notions (WCED 1987).

Two legitimate moral and social imperatives must be reconciled: the needs of the present and the needs of the future. The living standards of the billions today who do not enjoy the benefits of human progress – many of whom cannot satisfy even their basic needs – must be improved. At the same time, development patterns in both rich and poor countries must be altered in order to avoid leaving a bitter social and environmental legacy to future generations. For example, industrialization has relied on inexpensive and abundant fossil energy resources, particularly oil, natural gas, and coal. Yet, continued reliance on these fuels for the expansion of industrial activity risks committing the world to significant climatic alterations and extreme weather events for centuries to come (Nakićenović *et al.* 2000).

The profound challenge is to fashion a global development model that ensures rising standards of living without degrading the Earth's ecosystems, biodiversity, and climate. This requires a transformation in industrial processes, in the basis of modern lifestyles, and in the structure of economic development. A central theme of sustainability is harmonizing a complex and diverse set of goals that includes economic development, environmental preservation, and social justice. This coupling across issues and sectors is at the heart of the idea of sustainable development, suggesting that we may need to achieve these goals together, or not at all.

The implications are many. For policy, an integrated framework is needed to reflect the linkages between issues. For knowledge, a systemic perspective must complement specialization and reductionism. For popular values, a greater sense of connectedness to, and responsibility for, the human family, the broader community of life and the future must be cultivated.

There are those who downplay concerns about sustainability. Many hold philosophic objections to grand attempts to understand and guide human destiny, placing their faith in the capacity of the free market, human ingenuity and a homeostatic biosphere to provide timely responses to environmental and resource pressures. This worldview suggests minimalism toward development and environment policies, beyond steps to get competitive and maximally unfettered markets to function at local, national, and global scales. Indeed, this perspective is ascendant in many arenas, especially where economists of the neoclassical school advise governments and formulate policy (Beckerman 1995).

From the perspective of many ecologists and adherents of the sustainable development paradigm, this emphasis seems dangerously naive. The risks of relying on market and natural responses to correct perilous tendencies – and being wrong – are huge. The adoption of proactive policies and actions to avoid risks of ecological breakdown, resource degradation and related social friction appears the only prudent course under conditions of such uncertainty.

A major problem in joining these frameworks is the incommensurability of monetary costs as defined by markets, and environmental costs which are often long-term, multidimensional, and inherently normative (e.g., the cost of an "excess death" or of a lost species). There is no consensus, or even compelling methodology, for comparing the costs of climate change, for example, to the monetary costs of preventing it. Furthermore, there are great difficulties in embracing the interests of future generations – who cannot "vote" in today's market place – with the immediate bottom-line concerns of today's producers and consumers.

A source of hope is the growing realization that, once a sufficient standard of living has been reached, quality of life can expand without parallel increases in material requirements. The consumer society and its presumption of ever expanding material wants, is not synonymous with development and greater human welfare. Sustainable economic development can be based on qualitative expansion – growth in knowledge, in human capabilities, in social capital – that does not imply ever-increasing material wealth and environmental pressure.

Ultimately, the notion of sustaining the planet itself is a value that cannot be derived from economic doctrine. For those who adopt this value, a minimum requirement for policy will be to reduce the risk of undermining the conditions for human opportunity and activity in the future. We are at the early stages of operationalizing sustainability as a practical basis for action. This will require defining sustainability targets, laying out development scenarios that conform to those targets, and fashioning policy strategies for achieving goals.

The socio-ecological system

The evolving world system can be considered a *socio-ecological system*, comprised of environmental and human subsystems and their interactions (Gallopín *et al.* 1989, Shaw *et al.* 1991) as illustrated in Figure 1.1. The environmental subsystem, in turn, is composed of

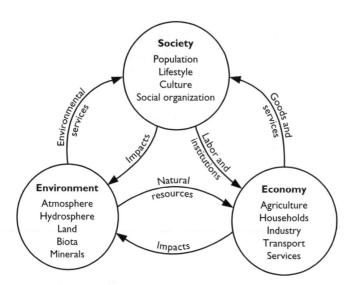

Figure 1.1 The socio-ecological system.

ecosystems, biophysical processes and other aspects of the natural world. The economic system includes capital, labor, other inputs, and the production processes in which they are used. The social subsystem includes consumption patterns, demographics, and culture. Subsystems interact in numerous ways and are mutually conditioning, so that a sharp distinction between dependent and independent variables is not meaningful. However, at a given time, certain processes may dominate the dynamics of the whole system and, thus, may be considered "driving forces."

Socio-ecological systems can be defined at local, national, regional, and global scales. The global system is decomposable into a nested hierarchy of interacting subsystems at each of these levels. Their evolution is subject to a set of unique forces acting at the level of the spatial unit. They are also conditioned through a complex set of horizontal and vertical linkages. A rigorous analysis of development prospects at the national level, for example, would need to be mindful of significant interactions and influences with other nations. At the same time, decomposition into local regions is often required to shed light on water issues, land change, social tension, and other problems. Finally, a defining feature of the current era is that the dynamics of the global system as a whole increasingly affect the component subsystems. Regional socio-ecological systems interact through processes operating at the global level, such as cultural influence, environmental impacts, transnational institutions, trade, global governance, geopolitics, and migration (Figure 1.2).

In the broadest sense, sustainability refers to the capacity for socio-ecological systems to persist unimpaired into the future. This by no means implies stasis – an impossibility in complex and dynamic systems – but rather, the capacity to adapt and develop. A sustainable system is *resilient* in the face of extreme perturbations and *flexible* in responding to changing circumstances. Sustainability is a *process* of development, not a final state. While it is difficult to define that process precisely, it is less difficult to identify *unsustainability*, patterns that place the socio-ecological system at risk of degradation, instability, and collapse.

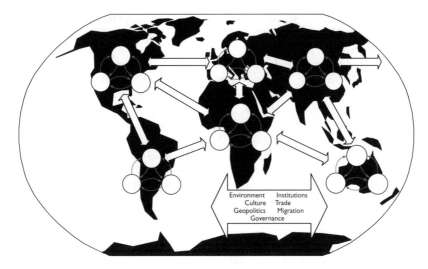

Figure 1.2 Global linkages.

The development of analytical methods and tools for assessing the sustainability or unsustainability of socio-ecological systems, and of approaches to achieving sustainability, is a daunting scientific challenge. It is a challenge that must be met if humanity is to successfully navigate the obstacles to a prosperous, just, and environmentally healthy future. Yet, analysis of economic, social, and ecological systems remains dominated by the reductionist approach that has proven so successful in enlarging our understanding of the world and of human nature. But conventional approaches and mindsets are not adequate in themselves for addressing the issue of sustainability in socio-ecological systems. Viewing the problem through narrow disciplinary, sectoral or thematic perspectives sheds light on the parts, but is opaque to the systemic character of the problem. Indeed, actions based on the results of compartmentalized analyses have in some cases aggravated environmental and developmental problems (Holling 1978, Myers 1984). The reductionist program of maximal decomposition of complex systems into component parts has been a powerful and effective force in modern science. But the strong interdependencies within socio-ecological systems and the ontological irreducibility of the whole require that a systemic approach complement the conventional analytic paradigm.

The new situation

The world system has entered an era of increasing complexity with many novel aspects (Munn *et al.* 1999). The new science of complex systems shows that unpredictability and surprise lie deep in the fabric of reality, not only at the microscopic level where the Heisenberg uncertainty principle applies, but also at the macroscopic level. In particular, the behavior of socio-ecological systems is indeterminate at critical branching points where the possibilities for further evolution proliferate. Where policy typically has reacted to well-understood problems, sometimes supported by scientific analysis and prediction, the inherent uncertainty of complex systems changes the burden on both science and policy. An adequate approach to sustainability would emphasize scientific uncertainty, the risk of destabilizing surprises, and the need to avoid them through the application of precautionary policy principles.

These ontological and epistemological characteristics of complex socio-ecological systems have implications for decision-making processes, as well. The critical problems of sustainable development are generally laced with both high scientific uncertainty and normative content. Whether considering actions for climate change mitigation, ecosystem preservation or global trade rules, policy formulation is debated in a context of scientific imprecision, the need to avoid unpleasant surprises, and the clamor of voices representing diverse interests. In response, decision-making processes are becoming more participatory and inclusive of affected stakeholders, and less technocratic and authoritarian. The legitimation of diverse criteria such as environment, human rights, gender, and security, and the emergence of new social actors such as global governance mechanisms, transnational corporations, and non-governmental organizations, increase the complexity of issues and solutions.

The 1992 Earth Summit marked a recognition that environmental, social, and economic concerns are closely connected and must be pursued jointly. Yet, structures of power and habits of mind change slowly. Many development efforts are still narrowly focused, and effective models for a more integrated approach to sustainable development are still lacking. Moreover, many individuals and institutions have a stake in preserving existing arrangements. Large landowners resist land reform, energy interests lobby for climate inaction, elites oppose commitments to eradicate poverty, and multinational corporations seek unfettered access to natural resources. In the absence of a widely shared awareness of the necessity for change and a positive alternative vision of the future, the inertia of complacency and vested interest restrains social transition.

The advocates of economic globalization argue that the rapid spread of a borderless economy is the path to prosperity for the global poor. Yet, the world remains stubbornly divided into rich and poor countries. While average income has doubled in developing countries over the past 25 years, the gap between their incomes and those in high-income countries has also doubled, as poverty persists, with about 1.2 million living in absolute poverty at less than $1 per day and nearly 3 billion at less than $2 per day (UNDP 2001).

Environmental and social links tie the destinies of rich and poor of the world more tightly than ever before in history. The environmental link is that rich and poor share the same planet. All are affected by the degradation of common global air, water, land, and biological resources. The social link is that poverty and political disintegration in the South can threaten the security and well-being of the North. If nothing else, the poor in a globalized world can export their misery, through migration, crime, terrorism, and disease. The coupling of destinies means that there are no separate solutions for South and North, East and West. It will take a truly global approach to reach a humane and sustainable future. Discussion and planning must move to the level of the human species, the biosphere, and global society.

The attack on the United States of 11 September 2001 by Islamic terrorists was one of those moments in history that define a "before" and "after." The event of 11 September 2001 was a cultural and political short-circuit that jolted many people from their complacency and caused many others to reconsider their priorities. It put the lie to the conceit that a superpower can withdraw from engagement in international affairs. The War on Terrorism is a geopolitical theme that will influence global affairs in the years ahead. But it could play out in very different ways. Certainly a cycle of violence and polarization could threaten global stability, liberal institutions, and development itself.

However, fundamentalist terrorism feeds off two contradictory impulses – a rejection of modernism itself by hard-core militants, and the anger of multitudes excluded from participation in modern economic and cultural life. If this is so, then a military and covert counterattack war may be necessary, but it is not sufficient. It must be complemented with a long-term strategic vision of an inclusive form of global development. The tragedy of 11 September 2001 and its aftermath provide a new context for understanding the link between human development, environmental stability, and national security. The new international alliances forged for mutual self-defense against a stateless and ubiquitous enemy could become the basis for cooperation on a broader agenda of pressing world problems.

The long-range future is open. No one can predict with any certitude how history will unfold in the twenty-first century. Yet, more than ever, we must consider the possible futures that might emanate from the turbulent conditions of our time, and the implications for human choice and action. In this spirit, we offer a perspective, framework and analysis of global possibilities as human civilization approaches critical branch points over the coming decades.

The human enterprise has begun a new evolutionary milestone – the planetary phase of civilization. This epochal transition poses weighty challenges for thought, policy, and action. To reflect on the human condition and our collective destiny in the early years of the new century is to enquire about the troubling perils ahead as we drift along the arc of history. But it is also to examine the possibility of a salutary passage to a more just and sustainable global society.

Chapter 2

Scenarios of the future

The essence of the quest for sustainability is a concern for the well-being of the long-range future. Speculation about ultimate human destiny is not new – it is expressed in most cultures through their mythology and religion. But in our time the question of the future has been mainstreamed, moving to the center of policy deliberation and the research agenda. To act wisely in the present, we must take a long view that considers the implications of our actions across a time horizon of several generations.

Uncertainty and the scenario approach

The challenge to science is to develop an adequate understanding of the dynamics of our increasingly complex socio-ecological systems, to illuminate future risks and uncertainties, and to inform policy responses. This requires the development and elaboration of techniques that are novel in the scientific setting. Analyzing human-environmental interdependencies over generational timescales cannot rely on methods that assume a strictly deterministic progression. Extrapolation based on present patterns in order to anticipate human affairs may be legitimate over the short term, but is it inadequate as time frames expand from months and years to decades and generations? Fundamental uncertainty is introduced both by our limited understanding of human and ecological processes, and by the intrinsic indeterminism of complex dynamic systems. Moreover, the future development trajectory will depend on human choices that are yet to be made.

Scenario analysis offers a way to consider long-range futures in light of the uncertainties inherent in socio-ecological systems and to examine the requirements for a transition to sustainability for such systems. Scenarios are stories about the future with a logical plot and narrative governing the manner in which events unfold (Cole 1981, Miles 1981, Schwartz 1991). Since they are not deterministic projections or forecasts, scenarios can examine the possible behavior of complex systems that may exhibit novel behavior, emergent properties and discontinuous transitions to new conditions.

Scenarios usually include images of the future – snapshots of the major features of interest at various points in time – and an account of the flow of events leading to such future conditions. Compelling scenarios need to be constructed with rigor, detail and creativity, and evaluated for plausibility, self-consistency, and sustainability. Scenario analysis challenges us to ponder critical issues and to explore the universe of possibilities for the future. Scenarios also clarify alternative worldviews and values, challenge conventional thinking, and encourage debate. Since they embody the perspectives of their creators, either explicitly or implicitly, they are never value-free. At their best, scenarios draw on both science – an understanding of historical patterns, current conditions and physical and social processes – and imagination, to conceive, articulate, and evaluate a range of plausible socio-ecological pathways (Raskin *et al.* 1996).

We wish to use scenario analysis to shed light on the problem of global sustainability. To that end, we consider scenarios that reflect contrasting social visions, identify significant causal processes that shape each story of the future, and highlight critical decision points. At such points, the choices of many actors and interests – individuals, corporations, labor, policy-makers, political institutions, cultural and spiritual leaders, and environmental activists – can significantly influence which global pathway emerges from the panoply of possible futures. Ultimately, the scenario approach can provide a common framework for diverse stakeholders to address the critical concerns of our time, and a forum for discussion and debate on the sustainability transition. In the near term, scenarios can offer guidance to the national and international policy community for converting the sustainability principle into practical policies and actions.

Ideally, scenarios would be (Swart 1996):

- *global*, with regional and, ultimately, subregional disaggregation;
- *comprehensive*, with integrated treatment of major environmental, social and economic issues, and interactions;
- *analytically rigorous*, with regard to use of data and scientific theory; and
- *diverse*, with representation of a range of future visions, values, and worldviews.

Scenario development is best approached as an ongoing process. Scenarios become more refined in the course of time, benefiting from the feedback of policy, environmental and public interest communities, better data and scientific advances. The process of refinement and the identification of limitations in available analysis provide useful guidance for prioritizing research agendas and data development efforts. In this iterative manner, the degree of spatial resolution, the level of thematic detail and the integration of scientific theory can evolve with time. The diversity criterion should be incorporated even in preliminary stages of scenario development, so that a rich range of perspectives can inform the analysis, and a process can be launched for communication, consensus and action.

Key elements in scenario formulation are illustrated in Figure 2.1. The development of scenarios generally begins with the characterization of the *current state*, which describes the initial conditions of the system under discussion in words and numbers (in the case here, the global system comprised eleven global regions). A careful specification of focal issues and key questions to be addressed is required to make the description of the current state tractable and useful. Depending on the context, the objectives of the scenario exercise might be to build awareness and stimulate the imagination, inform planning decisions or guide policy formulation.

With the focus and objective delineated, critical dimensions are formulated. They define the multidimensional space within which scenarios can be mapped or constructed. The dimensions are descriptors of important attributes of the images of the future. Examples of possible dimensions are economic growth, social progress, environmental quality, and conflict level. Often, a set of indicators is developed that provides a summary characterization of the scenario and a report card on its consequences for focal issues, such as environment, technology, and poverty.

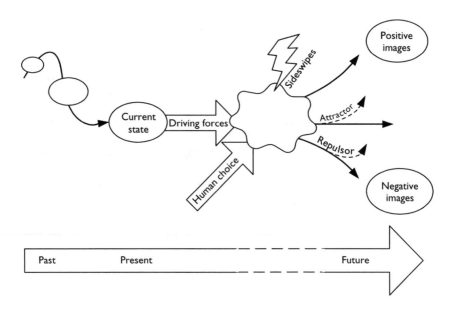

Figure 2.1 Scenario dynamics.

Next, the dynamics governing how the scenario unfolds must be specified. An essential step is the identification of the major *driving forces*; the factors, trends, and processes that drive the system forward from the current situation. As they unfold and interact, they condition the state of the system and the focal issues. Some driving forces are the same in all of the scenarios, especially slowly changing phenomena (such as human population growth and physical infrastructure), and processes already in the pipeline (the teenage population 15 years hence). Others represent critical uncertainties, the resolution of which would fundamentally alter the course of events, such as whether the world will continue to move toward a globalized economy, potential climate impacts, and the rise of terrorism.

All scenarios unfold according to an internal logic that links the elements into a coherent plot. The challenge is to identify narratives that capture the dynamics of the situation, while communicating effectively. A given set of driving forces is compatible with a range of plausible plots. Sophisticated long-range scenario exercises explore a small number of alternatives that bracket a broad range of qualitatively distinct outcomes.

The end point of the scenario is an *image of the future* – a description of the situation that results from the unfolding of the scenario. The image is a snapshot of conditions at some point in the future. In scenario *forecasting*, the image is a logical outcome of driving forces and the resolution of critical uncertainties as they evolve in the scenario narrative. The question of the forecaster is: Where are current trends taking us? By contrast, *backcasting* begins with a normative future condition; for example, a sustainable world in the year 2050. The questions become: Is there a plausible path to the future state? What human choices and actions are required to bring it about (or avoid it)?

The construction and interpretation of a scenario will be influenced by the beliefs and theoretical assumptions of the analyst, and others who have contributed to the scenario's formulation. The account of the mechanisms leading to alternative scenarios and judgment of the efficacy of alternative actions is guided by the worldviews of the analysts, although this is rarely made explicit (Herrera et al. 1976, Miles 1981, von Asselt and Rotmans 1997). Though always difficult, critical reflection and explication of the philosophical predisposition informing a scenario is an essential aspect of scenario description and documentation.

Long-range scenarios must recognize the role of deliberate human actions and choices in shaping the future. Human choice is influenced by cultural preferences and social visions, and by psychosocial factors that are not well-understood. As a complement to the driving forces, it is useful to introduce the concept of *attractors* or *repulsors* (Figure 2.1). These are visions of the future that can substantially redirect beliefs, behaviors, policies, and institutions toward some futures and away from others (Raskin et al. 1996). If the strength of the attraction or repulsion can be assumed to be weak, or if the actions they stimulate are consistent with current driving forces, the future evolves in what is sometimes referred to as a business-as-usual scenario. If, on the other hand, they are relatively strong, or act in a direction that is inconsistent with current driving forces, then future images can motivate intentional human actions for reaching desirable outcomes – or avoiding undesirable ones. One must imagine a future before acting rationally to create or avoid it. Subjective visions of future states, operating through human awareness, choice, and agency become objective forces conditioning the evolution of socio-ecological systems.

Also shown in Figure 2.1 are *sideswipes*, major surprises that can influence the future strongly – a world war, "miracle" technologies, an extreme natural disaster, a pandemic, or breakdown of the climate system.

Global scenarios in historical perspective

The exploration of the future has been a constant in human history, expressed through mythology, religion, social visions, and literature. But the modern era of global scenario building goes back only several decades. It was inspired by apocalyptic fears about the long-range fate of the planet. Initially, the Cold-War threat of nuclear holocaust stimulated strategic thinkers to engage in hypothetical war games and "thinking about the unthinkable" (Kahn 1962). Then, with the first flowering of global environmental awareness in the 1970s, scenarios turned to the new concern that population and economic growth, resource depletion, and pollution could lead to global social and economic collapse.

Early work included mathematical simulation models (Meadows et al. 1972, Mesarovic and Pestel 1974, Herrera et al. 1976), qualitative exercises (Kahn and Wiener 1967, Kahn et al. 1976), input–output analysis (Leontieff 1976), and eclectic approaches (Barney 1980). Reviews and critiques of global assessments introduced fresh insight (University of Sussex 1973, Meadows et al. 1982).

A second wave of global scenario studies have included narrative scans of alternative futures (Milbrath 1989, Burrows et al. 1991), an optimistic analysis by the Dutch Central Planning Bureau (Central Planning Bureau 1992), a pessimistic one by Kaplan (2000), a consideration of surprising futures (Svedin and Aniansson 1987, Toth et al. 1989), and a scan of possibilities from a business perspective (WBCSD 1997). The climate change issue gave rise to numerous model-based world energy scenarios, most importantly, those of the Intergovernmental Panel on Climate Change (IPCC 1992, Nakićenović et al. 2000). In addition, authors have revisited first-wave studies and affirmed their essential findings despite intense and sometimes rancorous criticism in the interim (Meadows et al. 1992, Barney 1993).

Methodological aspects

Useful scenario analysis for sustainability assessment must incorporate existing knowledge of ecological, economic, and social systems; allow for fundamental transitions and alterations in system dynamics; and examine a broad spectrum of plausible futures. Scenario studies that hope to inform policy and build public awareness must be transparent, allowing for effective critique, revision and participation by a broad spectrum of stakeholders.

From this perspective, at a deep epistemological level, the current generation of integrated assessment models has fundamental limitations for representing complex and open human systems. Many global studies have relied heavily on computer simulations with the desirable aim of consistency and replicability. The underlying metaphor is the global system as a complicated machine, which unfolds deterministically from initial conditions based on a set of equations that govern its mechanics. The fundamental structure of the system remains invariant. Futures differ because of quantitative changes of parameters (e.g., economic and demographic assumptions), rather than by transformations in basic relationships (e.g., alternative lifestyles, institutions, settlement patterns, and values). With this approach, model-based scenarios become merely sensitivity analyses for only one class of long-range possibilities – those that evolve gradually from current conditions and dominant development trends.

In general, simulation models are best at dealing with quantitative variables and relationships, where the structure of the system, as expressed through its functional relationships, is invariant. They require a well-developed level of knowledge about the values of the variables, parameters, and their relationships. Therefore, simulation models can capture only those elements that are reasonably well-understood, amenable to quantification and unfold smoothly and deterministically. Also, in many models, a high level of spatial aggregation masks the local specificity underlying calculated average global and regional trends.

Overzealous reliance on mathematical models has diminished the role of the underlying qualitative narrative of scenarios, so that the world behind the figures remains obscure. Important elements of the socio-ecological system that are non-quantifiable either in principle (e.g., culture) or in practice due to inadequacies in data or scientific theory have not been given appropriate weight.

This standard modeling strategy is, therefore, highly restrictive. The assumption of incremental and deterministic change does not adequately capture the inherent uncertainties in the global system and the capacity for surprise. In fact, global trajectories with qualitatively different behaviors and structures could plausibly emerge from the accelerating change and turbulence of globalization. For example, environmental and social stresses could spiral out of control, and the world could descend into conflict or authoritarianism. At the other extreme, new social visions and institutions could emerge. At critical *branch points*, the system can bifurcate in fundamentally different directions (Gallopín *et al.* 1997). Essential information about global possibilities – system shifts at critical moments, emergent phenomena, qualitative features such as culture – are not illuminated by mechanistic modeling.

How can we think about the future in a structured and rigorous way once we recognize that the mechanistic model is inadequate? Ultimately, it seems that a revolution in science will be needed that is analogous to the transition in physics from Newtonian to quantum mechanics. Where modern physics introduced the Heisenberg uncertainty principle, discontinuous state change and the irreducible role of the observer, a new sustainability science needs a conceptual framework that reflects the inherent uncertainty in global systems, the possibilities for abrupt changes and the importance of human actors in setting the dynamics of the system.

Recent methodological progress has involved unifying two parallel streams in the scenario literature; quantitative scenario analysis and narrative scenarios. This fusion can be found in recent work by the Global Scenario Group (GSG 2001) and the International Panel on Climate Change (Nakićenović et al. 2000). The challenge is to devise appropriate quantitative techniques to help tell the scenario stories (Raskin et al. 1999, 2002).

Narratives can provide carefully constructed stories of sequences of events that can lead to contrasting futures, while underscoring critical uncertainties and discontinuities. They are able to capture qualitative factors (values, culture, behaviors, and institutions), whole system shifts and surprise. But they lack the quantitative rigor and replicability of modeling analysis. The narrative approach must be fused with structured quantification in a way that transcends the mechanistic paradigm. Quantitative analysis can offer a degree of structure, discipline and rigor. Narrative can offer texture, richness and insight. The art is in the balance.

Timescales

It is important to distinguish between slow and fast dynamics operating within socio-ecological systems. High-level structures such as governance systems, economic modes of production, and cultural preferences tend to change relatively slowly. Environmental processes can also be slow; for example, significant climate response to human disturbances of the carbon cycle can take many decades. By contrast, the dynamics of lower subsystems can be fast, like the changing behavior of individual consumers in response to financial signals, technological innovation or the quality of local water resources in response to emissions.

The tension between the slower processes of the whole and the more rapid changes of subsystems shapes the critical uncertainties in the system. Over the short term, the slower processes contain the faster processes. But if a system becomes progressively more complex as it evolves, it can become more brittle and vulnerable to the influence of low-level, fast dynamics (Holling 1986). This can induce a rapid qualitative change to a new state. For example, the loss of a key species may precipitate the abrupt transformation of an ecosystem. In social and economic systems, the growing extent and speed of global communications is accelerating high-level processes, and injecting more potential for surprise into the global socio-ecological system (Gallopín 1991).

This discussion of the role of different timescales in the evolution of socio-ecological systems offers an additional perspective on the limitations of mechanistic models. An assumption of long-term persistence of the slow dynamics of large-scale structures – a key feature of conventional models – is associated with scenarios in which the future is shaped strongly by the past without abrupt and fundamental discontinuities. We refer to this condition as *structural invariance*. For the long term, however, the hypothesis of structural invariance has to be relaxed. As we consider the possible outcomes of the current era of planetary transformation, long-range global scenarios that go beyond the assumption of continuity of conventional structures must be considered.

Critical trends and driving forces

A number of important trends and forces are currently shaping global and regional developments (Raskin and Kemp-Benedict 2002). These megaprocesses set the initial direction and rate of change of globalization, shaping but not determining the future evolution of the development trajectory. Indeed, it is the manner in which human choice and complex socio-ecological systems respond to these combined forces that defines alternative futures.

In the discussion below, we group these trends and forces into major categories: demographics, economics, social and cultural change, technology, the environment, and governance. Of course, these factors are not independent of one another; but rather, are mutually conditioning components of a unitary process of global and regional change.

Demographics

Important demographic trends include population growth, the changing distribution of population between regions, urbanization, and shifting age structure in many countries.

It took over 200,000 years, from the first emergence of modern humans, for human population to reach 1 billion people, but only several hundred to reach 2 billion in the mid-nineteenth century. An additional 4 billion have been added since, with the sixth billion added in the last decade of the twentieth century. But the acceleration of population growth appears to have reached its peak. In the process of modernization, nations tend to undergo a demographic transition – population stasis first gives way to rapid growth as better nutrition and health reduce mortality rates, then growth slows as birth rates decline, eventually leading to population stabilization at higher levels. The transition is now making itself felt at the global scale, as the growth rate in aggregate world population begins to slow.

The fundamental premise of most long-range population projections is that national populations will eventually stabilize, with the time frame for the transition to stabilization the key variable. In all cases, however, the momentum behind population growth continues over the coming decades as life expectancies lengthen and the young in developing regions, who constitute a large portion of the population, reach childbearing age. Population patterns after about 2025 will depend on a number of factors, the most important of which is fertility. When the fertility rate exceeds about 2.1 children per woman – called the *replacement rate* – population will continue to grow. Currently, fertility rates are the highest in Africa, at nearly six children per woman, which if maintained would lead to a doubling in that region's population in 24 years, assuming present life expectancies. By contrast, China's fertility has dropped below the replacement rate, which if maintained – and this is a big uncertainty – will lead to a stabilization in that country's population within a generation. Fertility rates for other developing regions fall between these figures, while those for many developed countries are below the replacement rate.

Global population growth is expected to gradually stabilize over this century, but not before adding perhaps 3.5 billion people to the more than 6 billion that are alive today. Virtually all the additional population will be in developing countries, with modest growth in the United States and other rich countries expected from immigration rather than significant internal growth. Already at 80 percent, the fraction of world population in developing countries will continue to increase, with important implications for the environment and development. Rapid development and increased education, especially of girls, can accelerate the transition to slower population growth. But underdevelopment also can change trends, as revealed both dramatically and tragically by the HIV/AIDS pandemic in Africa.

The relationships among population growth, human development, and environmental degradation are highly varied and controversial (Raskin 1995). Some have stressed the primacy of population growth in undermining sustainable forms of development (Paddock and Paddock 1967, Ehrlich 1968, Brown 1978, Ehrlich and Ehrlich 1990) while, at the other extreme, "optimists" view the Earth's carrying capacity to be indefinitely expanded

Table 2.1 Urban populations (billions)

Region	1950	1995	2025 (projected)
World	0.8	2.6	4.8
Developing regions	0.3	1.6	3.6

Sources: WRI (1996b) and UN (2001).

through technological and institutional ingenuity (Kuznets 1967, Simon 1981). Others regard population as secondary in this context, stressing instead the large disparity among countries in average consumption and appropriation of resources (Chadwick 1994) – a "consumption gap" that continues to widen both between and within countries (UNDP 2001). Nevertheless, for a given set of socioeconomic development conditions, population growth increases the pressure on resources and the environment. This effect is most pronounced among the very rich, where each additional person accounts for huge resource and environmental investments, and the desperately poor, where the rationality of survival may imply the liquidation of natural capital to meet immediate needs.

Rapid urbanization is another major demographic trend. The world is in the midst of a massive transition from a predominately rural to a heavily urban society (see Table 2.1). The global urban population increased between 1950 and 1995 by more than a factor of 3. Over 2 billion more will be added to urban populations by 2025, mostly in developing regions.

The long urbanization trend has spawned huge "megacities," whose number and size have grown steadily. In 1950, there were two metropolitan areas with populations over 8 million – New York and London. Today, there are 25 – with all but five in developing countries – and nearly 400 urban agglomerations of over 1 million people (Brinkhoff 2001). Especially in poorer countries, urban planning institutions have been too weak to cope with rapid urban growth, which turns towns into cities, cities into megacities, and, if current trends continue, megacities into conurbations – continuous networks of urban centers. The deterioration of inner cities in some areas, and the growth of shantytowns on the periphery in others, undermine social cohesion, with high levels of social disparity, crime and violence. The elite adopt affluent urban lifestyles amidst a growing underclass living in squalor, often with inadequate sanitary, health and educational services. Moreover, urban tensions foster the suburbanization process and the rise of the automobile culture with its toll on land, the environment and the social fabric.

Another notable demographic trend is the aging populations of many countries. Lower fertility rates and greater longevity in rich countries are already increasing the average age, as the elderly account for a growing fraction of populations. This phenomenon will gradually appear everywhere if the projections of declining fertility rates prove accurate. Societies will need to adjust, as productive populations must support a progressively greater population of the elderly. At the same time, conventional notions of retirement may give way to extended working lives, where the needs and limitations of older workers are accommodated. The ways in which consumption and work patterns are resolved ultimately will influence the character of environmental problems.

Economics

The expansion and transformation of the world economy is another significant transnational process governing the evolution of the world system. Accelerated by advances in information technology and the growth of international trade agreements, the organization of production, and consumer and financial markets, is becoming progressively globalized. This is reflected in the surge in the level of cross-border trade and financial flows, in both absolute terms and as a share of total economic activity. For example, world output increased at just over 2 percent in the 1990s, while merchandise exports grew at about 7 percent (WTO 2001).

Three major trends will alter the political and economic landscape in the coming decades – the emergence of new national economic powers, the expansion of transnational corporations, and the formation of global economic governance processes. The economic legacy of the final decades of the twentieth century includes the extraordinary rise and then financial crisis of the "newly industrializing economies" of Asia; the collapse of the Soviet Union; and the "lost decade" of the 1990s for Africa and Latin America, during which output per capita declined in many countries.

The world economy has become more regionally pluralistic, with economic expansion in developing countries, Japan and the European Union. The economy of China could pass that of the United States in the next 20 years, with other Asian and Latin American countries becoming progressively more significant global players. Under typical economic projections, the size of the economies of developing countries taken in the aggregate in 2025 will be about the size of all industrial countries today. A number of factors will shape regional economic prospects. The debt burdens of many developing countries constrain growth, while the push toward greater economic integration will alter prospects in a number of regions.

Interacting with the emergence of new national centers is the second structural transition, the increasing role of transnational corporations. The growth of huge enterprises operating in a planetary marketplace is a natural extension of the growth dynamic inherent in competitive market systems. The expansion of modern infrastructure and stable legal frameworks in many developing countries has facilitated the globalization process. The revolution in communications technology, information processing, and transportation has vastly increased the capacity of transnationals to move facilities, products and people to their best advantage. At the same time, new trade agreements and the globalization of financial and currency markets challenge national protectionist restrictions.

In general, the world economic liberalization program advanced by the political leadership of most countries, large corporations and emerging global governance mechanisms is a dominant force that is driving economic change. We shall refer to it as the *conventional development paradigm*. But the major structural changes now enfolding should not blind us to important counter-trends that resist rapid globalization and advance various conceptions of national and regional diversity. An unusual coalition of forces resists the trend toward globalization and shifts in economic dominance, including nationally based economic interests, geopolitical isolationists, and social and environmental activists who raise non-trivial concerns about the impact of global competition on environmental protection and community stability. Finally, globalization, indeed modernism itself, is challenged in some countries by strains of religious fundamentalism, which find their most extreme expression in global terrorism against Western-dominated world development.

Social and cultural change

The explosive spread of information technology and the ubiquity of electronic media have accelerated global cultural change. In particular, American consumer culture is rapidly permeating many societies. The rise of a global consumerist culture – acquisitive, youth-oriented, and hedonistic – is both a result and a driver of economic globalization. At the same time, the forces of global cultural homogenization trigger reactions from those who seek to preserve cultural diversity and traditional prerogatives. While apparently contrary phenomena, the simultaneous advances of a single global marketplace, on the one hand, and nationalist, traditionalist, and religious movements, on the other, are dialectically inter-connected. Each poses important challenges to democratic institutions (Barber 1995). The hegemonic thrust of globalization, and the resistance to it, increase tensions between and within nations.

Globalized networks of terrorists, crime syndicates, and drug cartels are dark sides of the dominant tendencies toward market expansion and assimilation of cultures into a law-governed, materialist, and capitalist global culture. The capacity of a handful of violent, suicidal, and well-organized religious extremists to wreak havoc on an international scale was horrifically demonstrated in the terrorist attacks of 11 September 2001 and in the events that followed.

The question that has been asked in the United States is – Why do they hate us? The answers have fallen into two opposing categories – the rejection of modernism, and the lack of access to it. The first answer sees Islamist extremism as a reaction to such modern con-cepts as pluralism, religious tolerance, and secularism; a fear of being extinguished through assimilation. The second answer points to poverty and despair as the root cause; the failure of development to provide opportunity.

These contrasting explanations imply very different policy responses. If terrorism is a reaction against modernism, than dominant world forces cannot palliate it; rather, it must be eradicated. If its appeal is traced to a failure of development to equitably extend oppor-tunities to join the modernist project, then a long-term reconsideration of global develop-ment policy would be in order. In fact, both explanations are credible, and reflect two sides of the paradox of global development. Globalization's remarkable success in spreading Western-style development breeds fanatical resistance among ideologues. At the same time, its stunning failure to offer hope to billions creates fertile conditions for recruits and supporters.

Increasing inequality and persistent poverty characterize the contemporary global scene. As the world grows more affluent for some, life becomes more desperate for others left behind by global economic growth. Economic inequality between rich and poor nations is growing, as are disparities between the rich and poor within many nations. In the United States, for example, the distribution of family incomes has steadily widened over the past 30 years. Many families in the lower half have actually lost economic ground, while the incomes of those in the upper 20 percent have soared, with many becoming extravagantly wealthy. The income gap between developing and developed countries has increased over the same period; the poorest 10 percent of the world's population has less than 2 percent of the income of the richest 10 percent (UNDP 2001). Inequity could continue to increase in coming decades, both within countries, as welfare policies are weakened, and between coun-tries, if international mechanisms are not strengthened for wealth transfer or stimulation of the poorest economies. Widening equity gaps within a society aggravate poverty, threaten social cohesion and undermine development; widening gaps among nations motivate ille-gal immigration and social tension, complicating attempts to forge joint solutions to global problems.

Due to the combined effects of increasing income inequity and population growth, the number of people in absolute poverty has not decreased, even as average incomes have increased in most countries. Persistent poverty has consigned billions to deprivation and suffering. At the same time, the transition to market-driven development is eroding traditional support systems and norms, leading to considerable social dislocation, criminal activity, and even terrorism. In some regions, rampant infectious diseases are important social driving forces affecting development. Gender inequality in some regions also hampers social and economic progress.

Impoverishment and inequality are critical problems for poorer countries, but even the rich countries have significant pockets of poverty, inequity, and unaddressed human need. As the world grows more interconnected, global poverty affects all through immigration pressure, geopolitical instability, environmental degradation, and constraints on global economic opportunity.

Technology

Rapid technological change is transforming the structure of production, the nature of work and the use of leisure time. As noted above, information technology – computers, the Internet and telecommunications – is a catalyst for economic globalization. It will likely continue to impact the structure of production (down-sizing, just-in-time manufacturing), the nature of work (telecommuting, marketing and sales techniques) and leisure time (home shopping, interactive games, and media access).

Technology also has the potential to exacerbate tensions within and between societies who are excluded from its largess, or resent its implications for the preservation of traditional cultural values. This has been called the *digital divide* – the gap between the electronically connected affluent and the excluded billions at the bottom of the economic pyramid. A critical uncertainty for development is whether that divide will be bridged with innovative technology that addresses the needs and capacities of the poor, and a new kind of entrepreneurship that is targeted at a market of some 4 billion people, each with a small income, but a large market in the aggregate (Prahad and Hart 1999).

Biotechnology could have an array of significant effects on future society. The most mature is the use of genetically modified organisms for agriculture, but around the corner are vast possibilities for pharmaceuticals and, with the successful mapping of the human genome, identifying and preventing disease. The technology also raises numerous concerns. Environmental risks include the escape of bio-engineered genomic material into wild varieties with potential unintended consequences for ecosystems. Ethical dilemmas are epitomized by the possibility of genetic engineering of humans. Legal disputes have centered on "bio-piracy," where profits from genetic material accrue outside the source country. Finally, the technology could increase the dependency of developing countries on the international agro-industrial system. The first wave of genetically modified crops was marketed hard and dispensed rapidly, and the public reacted negatively. The potential of this technology rests ultimately on the degree to which governments and the private sector address concerns carefully, transparently, and gradually.

Lastly, the miniaturization of mechanics could fundamentally alter medicine and industrial processes. The ultimate stage in this direction would be *nanotechnology*, the engineering of computers, motors, and machines at the molecular level. While still in the early stages of research and development, these devices could significantly alter medical practices, material science and computer performance, and have many other applications. These technologies are a continuation of a century-long process of dematerialization, where progressively less material input is required per unit of product, and automation, where smart machines replace manual labor. They have the potential to diminish environmental pressure and reduce labor requirements through robotization. The latter, if not linked to a general scaling back of the average workweek, could radically reduce livelihood and employment opportunities. In general, these productivity-enhancing technologies could have a profound effect on future societies, with the potential for increasing wealth while eliminating drudgery and environmental pressure, or – if not coupled to other social and cultural changes – of enormous social displacement and new socio-ecological risks.

Environment

The degradation of the global environment is a significant driving force that will continue to shape development. There is a growing awareness that individual countries cannot insulate themselves from global environmental impacts and that collective action is required in changing the basis of global governance. The issues fall into two broad categories – the depletion of natural resources (the "source" problem) and the pollution of air, land, and water (the "sink" problem).

The most accessible and economic nonrenewable resources, such as minerals or energy resources, gradually are being depleted. Growing global demands eventually will require the more efficient use of these resources and the development of substitutes. This challenge has been met over the past 30 years – concerns with "limits to growth" notwithstanding – largely by aggressively exploring for additional reserves and developing more sophisticated techniques for exploitation. Nevertheless, current trends could lead to shortages and dislocations in strategic materials over the next decades; in particular, a re-emergence of oil shortages and, with them the potentially explosive geopolitics of oil (Raskin and Margolis 1995).

Of even greater concern is the depletion and degradation of the so-called renewable resources such as fresh water, marine resources, forests and other ecosystems. Resources are harvested at rates greater than they can be replenished. This over-exploitation is damaging the natural systems that sustain renewable resources. At the same time, pollution and climate change reduce the productivity and resilience of natural systems. The pressure on land-based resources has been driven largely by the growth of resource-hungry industrial economies. But it has been linked also to the growing populations of impoverished people who have few options other than over-exploiting these resources. Eventually, however, the depletion of resources undermines rural economies. The trends of environmental and human impoverishment are coupled and mutually reinforcing.

Rapidly growing urban areas in developing countries are growing sources of environmental health risks, subjecting populations to urban pollution hazards, shortages of both clean drinking water and sanitation, and exposure to air pollution and toxic materials. The number of people who lack access to clean drinking water is still growing, and most large developing country cities fail to meet World Health Organization standards for air quality.

On a global scale, worldwide emissions of carbon dioxide – a major greenhouse gas – are rising rapidly. The emissions are a consequence of growing use of coal, oil, and natural gas, and of the loss of forests and the carbon they sequester. Greenhouse gas emissions are rising rapidly, reflecting growing energy use in developed countries, and population growth and industrialization in many developing countries. Rather than stabilizing greenhouse gas concentrations in the atmosphere, which would require a reduction of emissions by at least half, the world is accelerating the threat of global climate change.

Governance

For much of the industrial era, governance has centered on the nation-state and hierarchically structured private and public organizations. The global transition has ushered in a dual process. On the one hand, authority is increasingly decentralized and, on the other, new forms of supranational, regional and global governance mechanisms are emerging. On both accounts, the traditional prerogatives of the state are challenged.

The process of decentralization of authority and greater individual autonomy takes many forms. On an individual level, this trend is noticeable in increased emphasis on "rights" – human rights, women's rights, and so on. In the private sector, it is reflected in the form of "flat" corporate structures and decentralized decision-making – even in the rise of entities that have no formal authority structure, such as the Internet. In the public sector, the trend is noticeable in the spread of democratic governments, in the devolution of governmental authority to smaller and more local units, in separatist movements and in the emergence of civil society as an important voice in decision-making, although in some countries corruption, autocracy and civil strife continue to undermine democratization.

Alongside the decentralization of authority are parallel trends in changing forms of governance that seek to regulate and manage regional and global issues. Greater regional integration, such as the European Union, is at various stages of implementation throughout the world. Global governance takes the form of such mechanisms as international trade and environmental agreements. At the same time, the new global connectivity has created the space for illegal and terrorist organizations to form and operate.

These trends, and the rise of many new actors from citizens' groups to global corporations, make governance an increasingly complex process. Global capital markets, for example, are not under the control of any government and can destabilize even major world currencies. Global communications networks – from CNN, to cellular satellite phones, to the Internet – convey information that is increasingly difficult for even determined governments to control. The growing strength of these global private sector entities is in marked contrast to the continuing weakness of global governance institutions. On local, national and global levels, the growing number and influence of non-governmental organizations and citizens' groups is, in part, due to their ability to provide information and services that governments do not, or cannot, provide.

The progressive de-coupling of decision-making by transnational enterprises from national agendas is an important trend posing challenges to the traditional prerogatives of the nation-state and the capacity for macro-economic intervention. As stateless corporations expand, with little allegiance to any country, the potential for political tensions rises significantly. Whether these tensions are resolved through a gradual balancing of global and national governance and regulatory structures, or whether they are the source of clashes and destabilization, will be an important sub-theme in the story of the twenty-first century.

Scenario framework

The steady evolution of current trends, however dominant they may be, is not inevitable. Existing trends are subject to numerous uncertainties and potential surprises. The character of future society will depend on how the forces driving these trends evolve, what policies are adopted and what choices human beings make. A transition toward a planetary phase of civilization has been launched, but not yet completed. In the coming period, social and cultural factors will play out on a rapidly changing stage, marked by technological, environmental,

and institutional transformation. Vastly different futures could crystallize from these turbulent conditions.

Scenario exercises must organize the bewildering zoo of possible futures into some kind
of taxonomy. A practical structure for organizing global scenarios must balance between
two competing considerations. The goal of analytic rigor invites an expansive range of scenario variations for exploring the full richness and texture of future possibilities.
Conversely, the desire to communicate findings to a wide audience of non-specialists dictates brevity and clarity, not to mention resource constraints.

Generally, scenario exercises rely on a very few stylized scenarios for illuminating key
issues, contrasting choices and uncertainties concerning the future. In typical policy-
oriented studies, a "mid-range" scenario is complemented by additional scenarios which are
generated by varying key driving forces – such as population, economic growth, and technological change – across high-low ranges. Our intention here is to introduce a framework
for transcending the practice of reducing the rich diversity of long-term possibilities to
mere variation in quantitative assumptions.

It is useful to classify scenarios within a two-tier hierarchy: *classes* based on fundamentally
different social visions, and *variants* reflecting a range of possible outcomes within each
class. This procedure highlights the plausible qualitative transitions in fundamental directions for society in light of today's driving forces, future uncertainties and the many critical
individual and collective decisions yet to be taken. We begin with three broad classes,
which we call *Conventional Worlds*, *Barbarization*, and *Great Transitions*. These are distinguished
by, respectively: essential continuity with current pattern; fundamental but undesirable
social change; and fundamental and favorable social transformation.

Conventional Worlds scenarios envision the global system of the twenty-first century evolving
without major surprises, sharp discontinuities, or fundamental transformations in the basis
for human civilization. The continued evolution, expansion and globalization of the dominant
values and socioeconomic relationships of industrial society shape the future. By contrast,
the *Barbarization* and *Great Transitions* scenario classes relax the notion of the long-term
continuity of dominant values and institutional arrangements. Indeed, these scenarios envision profound historical transformations over the next century in the fundamental organizing principles of society, perhaps as significant as the transition to settled agriculture and
the industrial revolution.

For each class we define two variants, for a total of six scenarios. Within *Conventional
Worlds*, the *Market Forces* variant incorporates mid-range population and development projections, and typical technological change assumptions. The *Policy Reform* scenario adds
strong, comprehensive and coordinated government action, as called for in many policy-
oriented discussions of sustainability, to achieve greater social equity and environmental
protection. In this variant, the political will evolves for strengthening management systems,
rapidly diffusing environmentally friendly technology and reducing poverty. Whatever
their differences, *Conventional Worlds* variants share the premises of the continuity of institutions and values, the rapid growth of the world economy and the convergence of global
regions toward the norms set by highly industrial countries. In the *Market Forces* variant,
the problem of resolving the social and environmental stress arising from global population
and economic growth is left to the self-correcting logic of competitive markets. In the *Policy
Reform* variant, sustainability is pursued as a proactive strategic priority.

Barbarization scenarios envision the grim possibility that the social, economic and moral
underpinnings of civilization deteriorate, as emerging problems overwhelm the coping
capacity of both markets and policy reforms. The *Breakdown* variant leads to unbridled

conflict, institutional disintegration, and economic collapse. The *Fortress World* variant features an authoritarian response to the threat of breakdown. Ensconced in protected enclaves, elites safeguard their privilege by controlling an impoverished majority and managing critical natural resources, while outside the fortress there is repression, environmental destruction, and misery.

Great Transitions scenarios explore visionary solutions to the sustainability challenge, including new socioeconomic arrangements and fundamental changes in values. These scenarios depict a transition to a society that preserves natural systems, provides high levels of welfare through material sufficiency and equitable distribution, and enjoys a strong sense of social solidarity. Population levels are stabilized at moderate levels, and material flows through the economy are radically reduced through lower consumerism and massive use of green technologies. The *Eco-communalism* variant incorporates the green vision of bio-regionalism, localism, face-to-face democracy, small technology, and economic autarky. The *New Sustainability Paradigm* variant shares some of these goals, but would seek to change the character of urban, industrial civilization rather than replace it, to build a more humane and equitable global civilization rather than retreat into localism.

Many alternative scenarios can be constructed as variations and blends of these pure cases. More nuanced scenarios would reflect regional variations and the possibilities of discontinuous jumps at critical points in the development trajectory. It should be stressed that elements that adumbrate each of the scenarios exist simultaneously in today's world. No real future will be a pure form, but rather an admixture of different social forms and ideologies, with some tendencies dominant. The idealized taxonomy provides a useful framework and point of departure for more detailed explorations. *Conventional Worlds* is where the standard policy discussion occurs. *Barbarization* lurks as a danger, the punishment imposed on future generations for unwarranted complacency today. *Great Transitions* offers idealistic alternatives; futures that may seem utopian, but are perhaps no less plausible than a sustainability transition without fundamental social transformation.

The scenario structure is summarized in Figure 2.2, with indicative sketches of the behavior over time for six descriptive variables: population growth, economic scale, environmental quality, socioeconomic equity, technological change, and degree of social and geopolitical conflict. The curves are intended as rough illustrations of the broad direction of possible patterns of change in the scenario.

We expand on the scenario descriptions in the following chapters, in which we develop detailed quantitative characterization of some of the scenario variants. Idealized and simplified scenario trajectories are illustrated in Figure 2.3 in a "space" defined by population along one axis and economic scale along the other. All scenarios depart from today's world, influenced by current driving forces. By the year 2100 – the end of each trajectory in the figure – the path has fractured into many possible future worlds due to different responses to critical conditions.

Conventional Worlds scenarios incorporate standard projections of population and economic growth. During the next century, population more than doubles and economic output increases by more than elevenfold as developing regions gradually converge toward socioeconomic patterns in rich countries. In *Barbarization-Fortress World*, global income grows very slowly as the few become much richer and the many get somewhat poorer. In *Breakdown*, world population peaks in the middle of the century before falling due to famine, disintegration of health institutions and warfare, while incomes begin to fall with the collapse of the economic and technological base. In *Great Transitions*, population increases slowly to 2050, as the demographic momentum built into today's conditions plays

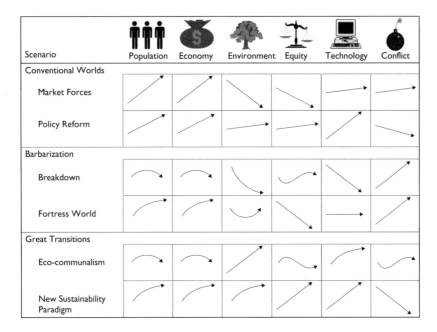

Figure 2.2 Scenario structure with illustrative patterns of change.

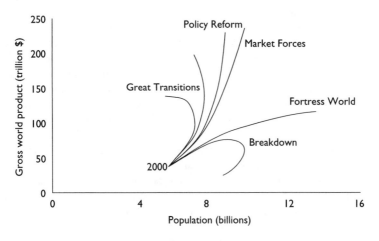

Figure 2.3 Scenario trajectories (2000–2100).

out, then stabilizes or gradually decreases over the rest of the century, due to assumed increased availability of contraception, increased status of women, and decreased desire for child labor. That this is plausible is demonstrated in the world today, where some industrialized countries have lower fertility rates than implied by even the lowest population growth levels shown in the figure.

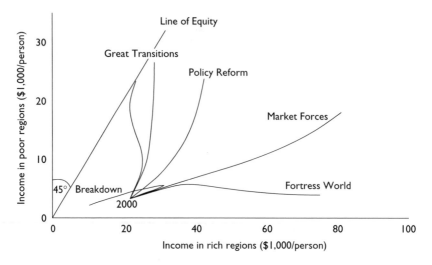

Figure 2.4 Equity patterns in the scenarios (2000–2100).

To get a better view of the equity implications of the scenarios, we plot the scenarios in an "equity space" (Figure 2.4). In this view, the axes are average incomes in rich and poor macro-regions (2000 figures are for Organization for Economic Co-operation and Development (OECD) and non-OECD, respectively). As a scenario approaches the *line of equity*, the difference between incomes in rich and poor areas diminishes. When a scenario moves parallel to the line of equity, the absolute difference in incomes between the regions is maintained. Despite relatively rapid annual percent growth of developing region economies assumed in standard projections, inequity persists in the *Conventional Worlds-Market Forces* scenario, as the curve moves away from the line of equity. In general, incomes gradually converge, in the sense that the ratio of average gross domestic product (GDP) per capita in regions to other regions decreases. However, the *absolute* difference increases as incomes soar in rich countries.

By contrast, the trajectory of the *Policy Reform* variant approaches the line of equity. Nevertheless, substantial income disparities remain, even by 2100, despite the assumed moderate income growth rates in rich regions, and substantial wealth transfers. In general, it is extremely difficult to redress fully the current global inequities under *Conventional Worlds* conditions that assume continuing economic growth in rich areas. *Barbarization-Breakdown* is a high-equity scenario but at low-income levels, while the *Barbarization-Fortress World*, by definition, is a case of extreme inequity. The *Great Transitions* scenarios approach equity conditions rapidly.

Alternative worldviews

The scenario framework offers insight on a range of contrasting possibilities for the global future. They address the prospects for an unprecedented phase of human history – the transition to a planetary phase of development – and in that sense they are visions with few antecedents in the history of ideas. At the same time, they are contemporary expressions of age-old social philosophies (Raskin *et al.* 2002). At the broadest level, the three scenario categories – *Conventional Worlds*, *Barbarization* and *Great Transitions* – correspond to three archetypal political sensibilities – the evolutionary, the catastrophic, and the transformative. These mindsets reflect different premises, values, and myths about how the world works. In the contemporary debate on the future, they expressed divergent outlooks on global prospects.

Evolutionists are optimists who trust that dominant driving forces and good governmental policy can deliver prosperity, social stability and ecological health. Catastrophists are pessimists who fear social and environmental crises will deepen with dire consequences for the future. Transformationalists share these fears, but believe global development offers opportunities for forging a new and better civilization. In a sense, these outlooks represent three different worlds – a world of incremental adjustment, a world of discontinuous cataclysm, and a world of structural shift and renewal.

The intellectual antecedents to the social visions embodied in our scenarios are summarized in Table 2.2. The *Market Forces* bias is that the hidden hand of well-functioning markets is the key to resolving social, economic, and environmental problems. An important philosophic

Table 2.2 Archetypal worldviews

Worldview	Antecedents	Philosophy	Motto
Conventional Worlds			
Market forces	Smith	Market optimism; hidden and enlightened hand	Don't worry, be happy
Policy reform	Keynes/ Brundtland	Policy stewardship	Growth, environment, equity through better technology and management
Barbarization			
Breakdown	Malthus	Existential gloom; population/ resource catastrophe	The end is coming
Fortress world	Hobbes	Social chaos; nasty nature of man	Order through strong leaders
Great Transitions			
Eco-communalism	Morris and social utopians	Pastoral romance; goodness of man; evil of industrialism	Small is beautiful
New sustainability paradigm	Mill	Sustainability as progressive global social evolution	Human solidarity, new values, the art of living
Muddling Through			
	Your brother-in-law (probably)	No grand philosophies	Que será, será

forerunner is Smith (1991), with contemporary representatives including many neo-classical economists and free-market enthusiasts. The *Policy Reform* orientation is that markets require strong policy guidance in order to avoid economic crisis, social conflict and environmental degradation. John Maynard Keynes (1936), writing at the time of the Great Depression, is an important philosophic predecessor to the belief that sustainable development requires proactive management of market globalization. The Brundtland Commission report (WCED 1987) was a seminal expression of this perspective, which underlies much of the official international discourse on environment and development.

The dark vision of *Breakdown* is that unbridled population and economic growth leads to a world calamity of ecological disaster, violent conflict, and the collapse of civilization. Thomas Malthus' (1983) grim calculus of geometrically increasing population outstripping arithmetically increasing food production was a precursor of this gloomy prognosis. This worldview surfaces in various forms in modern assessments of the global predicament (Ehrlich 1968, Meadows *et al*. 1972, Kaplan 2000). The *Fortress World*, which envisions authoritarian solutions to the threat of breakdown, is rooted in a pessimistic view of human nature and the skepticism about democratic solutions. It has a classic formulation in the philosophy of Hobbes (1977). Observers today need not be terribly pessimistic to fear that contemporary trends could drift toward a world of social polarization, haves and have-nots, and the withdrawal of elites to gated cities and secure enclaves; that is, some form of a *Fortress World*.

The *Eco-communalism* belief system of localism, human-scale societies, and enchantment with nature has links to the social utopians of the nineteenth century and the romantic reaction to industrialization (Thompson 1993), and the twentieth century small-is-beautiful philosophy of Schumacher (1972) and traditionalism of Gandhi (1993). With an added environmental emphasis, these philosophies live on in contemporary anarchistic visions of the good society (Bossel 1998, Sales 2000). The *New Sustainability Paradigm*, which seeks a humane and ecological form of global civilization, is the tacit and unformed view of nascent movements for a revised form of globalization. It has few historical precedents, although John Stuart Mill's (1998) speculations on the possibilities for a post-industrial society based on human development rather than material acquisition was prescient. The clarification of the new paradigm is a critical theoretical task for the future (Raskin *et al*. 2002).

This survey covers a range of historic and contemporary attitudes and philosophies. But many people subscribe to no coherent worldview, preferring to react to events as they arise. This is the *Muddling Through* bias shown in the last row of Table 2.2. Its practitioners may shun speculative philosophy as a matter of principle, they may be unaware of the global challenge, or they may simply not care. Whether this silent majority will remain complacent or become engaged actors is a key unknown in the drama of the global future.

Destiny and opportunity

The intellectual, moral, and cultural climate of the last decades has been dominated by a development paradigm obsessed with economic growth and globalization. A convenient faith that the rising tide of aggregate output would raise all boats justified indifference about the fate of the poor, as reflected in diminishing official development aid. By contrast, equity was a major international theme of the 1970s, and strategies for reducing the gap between rich and poor were at the top of the agenda (Steenbergen 1994).

Now political and cultural counter-trends challenge the conventional paradigm. Deeply concerned about detrimental impacts on communities, labor practices, and the environment, growing movements counter the dominant drift. Some accept the "anti-globalization" label as they seek to slow or reverse the momentum toward a global market system. Others argue for a fundamentally different character to global society that would privilege the issues of global justice, poverty reduction, and sustainability, displacing the fixation on economic growth.

Meanwhile, religious extremism fed by a complex ideological brew of hatred for infidels, fear of assimilation, and human desperation challenge the very possibility of an open global society. The events of 11 September 2001 and their aftermath will have repercussions that cannot be foreseen. The jolt of terrorism has shaken the world from its complacency and indifference, but the long-term implications and consequences are not clear.

The so-called War on Terrorism that has been launched in the aftermath of the attack may play out in renewed efforts to accelerate *Market Forces* and the incorporation of all countries into the global nexus. Alternatively, the new international cooperation, and growing recognition that pervasive poverty and environmental deterioration compromise international security, could feed a *Policy Reform* future. More disturbingly, an escalating cycle of violence, retaliation, and anger could lead to chronic global instability and conflict, and trigger a *Barbarization* scenario. But the potential for such a disastrous scenario could lead to a more hopeful future as well. The glimpse of the abyss and the exposure of the deep deficiencies of conventional development strategies could catalyze the search for a new paradigm – perhaps a *Great Transition*.

Chapter 3

Sustainability goals

The nations of the world pronounced their commitment to sustainable development at the 1992 Earth Summit in Rio de Janeiro and laid out a broad strategic agenda (UNCED 1992). While governments have yet to convert the rhetoric into action, Rio was a watershed that opened a new era of debate on the problem of development.

In broad brush, sustainable development seeks to harmonize human development with preservation of the environment. It is about reconciling the needs of the present generation with those of its descendents. At the heart of the concept is the powerful ethical imperative that we strive to consume and produce in a manner that is tempered by concern for the well-being of our descendants and the resilience of natural systems, while meeting the needs of people today.

These philosophical principles are compatible with diverse views on the specific implications of sustainability. No formula can give the optimal balance between the needs of people today, of future generations, and of the wider community of life. This is a normative issue that will be adjudged differently by various communities and individuals. While there can be no consensus on a blueprint for sustainability, the goals for the future in much of the sustainability literature center around such common themes as:

- eradicating absolute poverty and hunger;
- providing universal entitlement to basic social services such as health care and education;
- improving quality of life everywhere and expanding possibilities for fulfillment;
- reducing economic and social disparities;
- increasing environmental quality, with preservation of biological resources and eco-systems, decreased pollution, and climate stability;
- diminishing violence and armed conflict; and
- stabilizing global population.

The goals are closely coupled. For example, global populations will stabilize to the degree that access to shelter, food, health care, security, and education reach the poor. Also, poverty reduction and greater equity almost certainly are essential for preserving biological resources. Moreover, preventing increased violence and conflict may depend both on rising incomes and on increased equity within countries and between rich and poor regions.

To design and test appropriate policies, it is useful to express such broad goals in terms of specific quantitative objectives (Raskin *et al*. 1998). Of course, any set of criteria for a sustainable world must necessarily be tentative and partial, and subject to revision in light of new knowledge and political priorities. Nonetheless, a preliminary set of minimum objectives – both social and environmental – are needed to assess the challenge, guide actions and gauge progress.

Dimensions of resilience

What sustainable development seeks to sustain is the integrity and resilience of combined human and natural systems as they interact and condition one another over time. The sustainability of socio-ecological systems is a dynamic process of development, not a static condition. As a quality of coupled environmental and human systems, sustainability has both biophysical and socioeconomic dimensions.

The biophysical dimension

Biophysical sustainability seeks to preserve the functional and structural integrity of ecosystems, bio-geochemical cycles, and the natural resource base. These are sustainable insofar as they can support human material well-being in perpetuity, provide ecological services and preserve the natural heritage for human appreciation. Beyond these anthropocentric requirements, for many, the preservation of nature and the survival of species are intrinsic values that do not require utilitarian justification.

Development will be sustainable in a biophysical sense when human activity does not destroy the regenerative capacity of natural capital or irreversibly stress atmospheric, hydrological, or terrestrial ecosystems with waste and pollution. Sustainable development focuses on reducing the *throughput* – the flows of materials and energy into, and waste out of, production and consumption activities – toward levels that are compatible with renewable resource flows and assimilative capacities of ecosystems.

Throughput levels, in turn, are dependent on consumption patterns, population levels, production technologies, land-use management and other factors that determine the requirements for virgin materials and pollution loads. In one sense, the problem of sustainability is the conflict between rising throughput rates, driven by growing economies, and the finite capacity of the biosphere to provide resources and tolerate stress.

Sustainability implies living on natural "interest," not drawing down natural capital. Thus, it draws attention to the scale of the economic system and the degree to which human activity transforms nature and threatens environmental systems. Endless quantitative growth is impossible on a finite planet, although qualitative improvement in products, skills, and culture need never cease in an environmentally sustainable world (Daly 1996).

The socioeconomic dimension

The second face of sustainable development is socioeconomic. For many who have joined the sustainability discussion, though certainly not all, achieving basic social goals – the "needs" of today and of future generations in the Brundtland formulation – is taken as a pre-analytic moral imperative. Eliminating hunger, providing access to education and health services, and diminishing inequality between and within regions are sufficient goals on ethical grounds alone, whether or not they are necessary antecedents to environmental sustainability. Indeed, there are plausible "solutions" to the problem of environmental sustainability that are politically and socially repressive (Gallopín *et al.* 1997).

However, beyond the moral imperative for reducing human deprivation, there are important objective links that couple social goals to environmental conditions. For example, poverty is both a cause and an effect of environmental degradation (World Bank 1996). The desperately poor are likely to mine nature for immediate survival, not preserve it for a dubious future. Moreover, the social cohesion required for a comprehensive sustainability transition is undermined in a society where the needs of its citizens for material well-being and just treatment are not met. At the global level, regional disparities foster migration pressure, environmentally unfriendly trade and development patterns, and difficulty in negotiating international environmental agreements.

Thus, the socioeconomic and environmental aspects of sustainability are highly interdependent. If a society permits excessive environmental deterioration, it risks undermining the economic welfare of its citizens, the legitimacy of its political systems, and the endurance of its institutions. If a society festers with social tension and instability, it is not likely to make the environment a priority, or enjoy the institutional capacity for implementing a sustainable form of development. We take as a fundamental principle, therefore, that the socioeconomic and biophysical dimensions of the sustainability transition must be treated in a unified framework.

The notion of social sustainability calls attention to the stability, quality of life, and social cohesion in society. To the degree that distributional equity, political participation, and access to education, health and cultural services are perceived to be acceptable, a social system will enjoy the commitment, loyalty, and affiliation of its participants, and be better prepared to respond to changing endogenous and exogenous circumstances. At the other extreme, a system which is inequitable and coercive tends to be more rigid, prone to conflict, and less able to adapt gently to internal or external disturbances.

Finally, economic development is a precondition for a transition to sustainability. The wide adoption of sustainability principles will require that economic systems and distribution patterns provide basic human needs, reliable livelihoods, and freedom from drudgery. However, unlike the conventional focus of development programs, from a sustainable development perspective, both rich and poor countries face a development challenge. In rich countries and communities, the challenge is to transform the model of progress from ever-increasing growth in consumption to a culture of material sufficiency. Development would be measured by the growth of quality values – for example, through building of stronger community ties, engaging in meaningful leisure activities, and relating to nature.

In general, the concept of socioeconomic *development*, the expansion or realization of potentialities, must be distinguished from economic *growth*, or material accretion (Goodland *et al.* 1992). Material growth, the hallmark of the industrial era, is not indefinitely maintainable. Human cultural, intellectual, artistic, social, and technological development, together with the provision for basic physical needs, is not only compatible with sustainability, but also essential for its realization.

Regionalization

We are concerned here with global patterns and future possibilities. But the global system has many heterogeneous elements. It is a complex hierarchy of distinguishable *subsystems* at local, national, and regional levels. While each of these entities is partially separable and quasi-autonomous, the defining feature of the planetary phase of development is that global-level dynamics increasingly constrain and influence the joint evolution of subglobal units.

In global assessments, a balance must be struck between the desire for detail and disaggregation, and the goal of analytic coherence. At one extreme, global scenarios might be constructed with considerable spatial detail and attention to the interplay between local and global dynamics. At the other extreme, the world system might be treated as a single aggregate entity, which would allow only the crudest analysis of patterns of change. In practice, the degree of global disaggregation is limited by the practical consideration of data availability and, in the interests of clarity of communication, by the desire to avoid overwhelming detail.

There are many alternative ways of defining global regions; for example, by dominant religio-cultural practices, by agro-ecological zones, by river basin, and by socioeconomic system. Furthermore, there is never a sharp demarcation between regions, so certain countries can arguably be moved from one region to another without doing violence to the analysis. No configurations are without conceptual complications and daunting data problems. For example, the anachronism of defining the Former Soviet Union (FSU) as a region is necessary for now because that is the way data have been organized historically.

In this analysis, we have grouped countries into eleven global regions, based on the comparability of socioeconomic development and geopolitical considerations. The spatial structure for this long-range global assessment provides enough resolution for exploring important global variations and trade patterns, while not exceeding the availability of data and the capacity to grasp the main contours of the global system. The regional groupings and the countries included in each are displayed in Figure 3.1.

The analysis is ultimately constructed from country-level databases and assumptions. For clarity, scenario results for the eleven regions will be collapsed at times into "macro-regions": *OECD* (the high-income countries of the Organization of Economic Cooperation and Development consisting essentially of our North America, Western Europe, and Pacific OECD regions) and *non-OECD*. The non-OECD macro-region is, in turn, sometimes disaggregated into *Transitional* (FSU and Eastern Europe) and *Developing* (Africa, Latin America, Middle East, China+, South Asia, and Southeast Asia) macro-regions.

The discussion will refer to quantitative results for selected scenarios, which are grouped by region and macro-region and collected in the Annex in a series of tables and graphs.

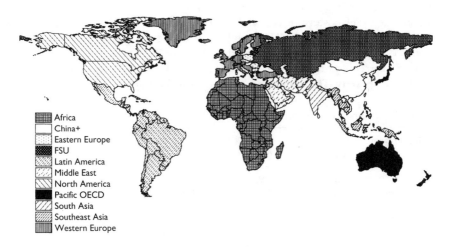

Africa
China+
Eastern Europe
FSU
Latin America
Middle East
North America
Pacific OECD
South Asia
Southeast Asia
Western Europe

Figure 3.1 The eleven regions.

Minimum targets

To establish quantitative objectives, three sets of choices must be made: first, which indicators are to be used to measure progress toward sustainability; second, which values of these indicators represent sustainable conditions and, hence, provide targets against which to measure progress; and third, how rapidly these targets are to be achieved. These choices are to some degree subjective by nature, dependent not only on the interpretation of uncertain scientific information, but also on the cultural preferences and interests of an individual, a community, a country. This implies that different societies might choose different sustainability criteria.

We consider below core global indicators and targets. They provide provisional guidelines for the scope, scale, and timing of strategic actions required for a transition to sustainability. However, there are no blueprints – sustainability indicators and targets will need to be refined over time in light of new information, events, and perspectives. Sound sustainable development practices will be an instance of *adaptive management*.

Social objectives

Taken broadly, the notion of setting *social goals* could be construed to imply fundamental normative questions, such as visions of the good society of the future. There can be no easy consensus on these matters, since the range of perspectives will vary across the full range of worldviews and political philosophies. The discussion here is more limited in scope, focusing on a narrower set of goals that have wide international consensus: the provision of basic human needs, such as adequate food, clean drinking water, and access to health care and education.

At a minimum, this report argues, sustainability must mean achieving such social goals. However measured, the human condition today falls far short of this standard. It is a sad legacy of our time that the extraordinary expansion of the aggregate global economy has not diminished the sum total of human misery. The percentage of people destitute may have declined, but not the absolute number. Not only has absolute poverty coexisted with rapid economic growth, to a large degree it has been generated by the very systems that provide the engine of that growth (Gallopín 1994).

The prevalence of poverty in the world today is an oft-told story that can only be summarized here (UNDP 1997). We begin by considering the stunning contrast between the highly developed OECD countries and the other non-OECD countries (Table 3.1).

Nearly 85 percent of world population reside in non-OECD countries, where 25 percent live in absolute poverty. The absolute poverty level is defined by the World Bank (2001) as

Table 3.1 Income, population, and poverty in 1995

Macro-region	Income ($ per capita)	Population (millions)	Absolute poverty (millions)
OECD	21,000	920	10
Non-OECD	3,240	4,750	1,200
World	6,120	5,670	1,210

Sources: Income from WRI (1998); population from UN (1999); non-OECD poverty figures from World Bank (2001); OECD poverty figures equated with hunger estimates.

individual consumption expenditures less than $1/day or about $370/year in 1993 dollars. The 1.2 billion living in such dire poverty exceeds the entire population of the OECD countries. Using GDP per capita as a proxy for average income, we see from Table 3.1 that people in OECD countries are 6.5 times richer than those in non-OECD countries. In making these comparisons, and throughout this report, national currencies are converted to common units using the purchasing power parity (PPP). National PPP-adjusted GDP per capita is taken from WRI (1998). The PPP approach, in which prices of a common "basket of goods" are compared across countries, gives a more realistic picture of relative incomes than the more commonly used market exchange rates (MER) (WRI 1996a).

When national GDPs are compared using PPP, rather than MER, the estimated incomes in most developing countries relative to rich countries are generally higher. In MER-converted terms, the ratio between OECD and non-OECD GDP per capita is about 19, considerably higher than the figure of 6.5 reported in the text. The 16 percent of world population that reside in the wealthier countries claim 56 percent of global income (the figure is 79 percent when GDP is expressed in MER rather than PPP terms). Moreover, the disparity between the world's rich and poor has been increasing (UNDP 2001).

There are many possible measures of the status of human development. Four indicators that represent key categories of human well-being are: chronic undernourishment (represented by *hunger* levels); availability of clean water (measured in terms of the *population with unsafe drinking water*); education (gauged by the level of *adult illiteracy*); and human health (a proxy is *life expectancy at birth*). The global situation for the four indicators is reported in Table 3.2. The indicators can be expanded and combined; for example, the Human Development Index (UNDP 2001) combines average income per capita, literacy, and life span.

Not surprisingly, the disparities that were observed in the incidence of poverty between OECD and non-OECD regions are also reflected in these more tangible measures. In non-OECD regions, over 800 million people are undernourished today, 18 percent of the population in those regions. Moreover, 1.35 billion people, roughly 28 percent of the population, do not enjoy reliable and sanitary sources of drinking water. The cumulative effects of the lack of basic education are indicated by the 29 percent of the adult population that is illiterate. Finally, life expectancy at birth in non-OECD regions remains substantially below OECD averages, though there have been impressive increases in recent decades. Life expectancy in developing countries rose from 46 to 62 years between 1960 and 1994, as infant mortality declined from 149 to 39 per 1,000 live births (UNDP 1997), and death from infectious diseases declined (WHO 1997b).

Values for the social indicators by region are presented in Table 3.3. For context, also shown in Table 3.3 are GDP per capita and a measure of income inequality (the ratio of the income of the poorest 20 percent to that of the richest 20 percent). Regional income inequality is computed as the population-weighted average of country-level data. Africa fares least well across all four indicators. More than one-third of the population is undernourished and

Table 3.2 Selected social indicators in 1995

Macro-region	Hunger (%)	Unsafe water (%)	Illiteracy (%)	Life expectancy (years)
OECD	1	1	2	77
Non-OECD	17	28	29	64
World	14	24	24	66

Table 3.3 Selected indicators by region in 1995

Region	Hunger (%)	Unsafe water (%)	Illiteracy (%)	Life, expectancy (years)	GDP per capita ($)	Poorest 20% divided by richest 20%
Africa	27	49	45	55	1,970	0.11
China+	14	20	18	69	2,960	0.12
Latin America	11	17	14	69	5,830	0.05
Middle East	15	27	38	65	7,330	0.11
South Asia	23	36	51	62	1,520	0.15
Southeast Asia	11	32	13	65	5,563	0.13
Eastern Europe	1	7	0	71	5,600	0.22
FSU	7	8	0	67	3,750	0.16
North America	2	0	0	77	26,470	0.10
Pacific OECD	0	0	0	79	21,320	0.26
Western Europe	1	1	4	76	17,400	0.20

Sources: Hunger (see Annex note S-2); unsafe water = 100 percent – access to safe drinking water; access to safe drinking water for 1980–95 (World Bank 1997, WHO 1997b), set to 100 percent for high-income OECD countries, FSU data partial; illiteracy = 100 percent – adult literacy, adult literacy for 1990–95 (World Bank 1997); life expectancy at birth for 1990–95 (World Bank 1997); GDP per capita in PPP (WRI 1998); income ratios (see Annex note S-1).

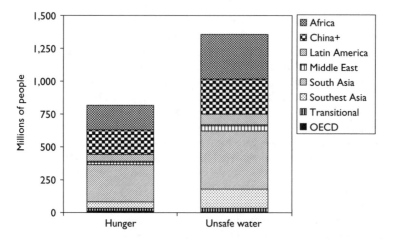

Figure 3.2 Population suffering from hunger and a lack of safe drinking water in 1995.
Source: See Table 3.3.

nearly half are without safe drinking water. South Asia, Southeast Asia and the Middle East also register high levels of deprivation, with China+ and Latin America not much better off. The information on hunger and unsafe drinking water is presented graphically in Figure 3.2.

Poverty levels in a given country depend on both average income and distribution of income. As a general principle, for a given average level of income, the more skewed the distribution, the greater the fraction of the population in absolute poverty. This relationship can be gleaned from the data in Table 3.3. Eastern Europe and Latin America have similar average incomes, but income disparities are much greater in Latin America. So are levels of hunger. Income distributions are similar in China+ and Africa, but average incomes in Africa are much lower than in China+, with correspondingly higher levels of absolute poverty.

Social goals for sustainability may be expressed in the language of reduction targets for each measure of human deprivation. Here we rely on the work of a series of international conferences that over the past few years have attempted to define goals for the reduction of human deprivation and the provision of the opportunity for a dignified life for all (BSD 1998). Each of the indicators reflects a different dimension of poverty.

At the 1996 World Food Summit, it was resolved that undernutrition was to be halved by the year 2015 (FAO 1996b). To achieve this goal, the number of undernourished people must decline from 820 million in 1995 to roughly 410 million over 20 years. Based on the historical record, this is quite an ambitious goal, because the number undernourished fell only about 70 million between 1970 and 1990 (FAO 1996c).

In the spirit of the World Food Summit goal, but allowing for some slippage, this report proposes a minimum target for hunger reduction in *Policy Reform* scenarios of halving undernourishment by 2025 and halving it again by 2050. The targets are defined in terms of the absolute number of undernourished: by 2050, the number of undernourished people is targeted to fall to one-quarter of its 1995 value (Table 3.4). When expressed as a fraction of world population rather than absolute numbers, hunger falls faster, from about 14 percent in 1995 to just over 2 percent in 2050, since population is assumed to grow from about 5.7 billion in 1995 to 8.9 billion over this period, assuming mid-range population projection levels (UN 1999).

Strong commitments to universal provision of safe drinking water have been a recurrent theme in human development conferences and in the policy goals of multinational organizations (WHO 1997a). Unfortunately, there is a history of bold pronouncements that go unmet, such as the Decade of Safe Drinking Water in the 1980s. As with hunger, we assume that the population without safe water is cut in half by 2025. Applying this target separately to urban and rural areas, this corresponds to a minimum level of safe water service of about 85 percent for rural and 95 percent for urban dwellers. The goal for 2025

Table 3.4 Global social targets

Indicator	1995	2025	2050
Hunger			
Millions of people	820	410	205
% of 1995 value	—	50	25
% of population	14	5	2
Unsafe water			
Millions of people	1,360	680	340
% of 1995 value	—	50	25
% of population	24	9	4
Illiteracy			
Millions of people	1,380	690	345
% of 1995 value	—	50	25
% of population	24	9	4
Life expectancy (years)	66	>70 in all countries	

may be compared to today's figures of 50 and 75 percent safe water access in rural and urban areas, respectively, in many developing countries. The 2050 target assumes another halving, corresponding to over 95 percent access to safe water everywhere.

Similarly, the World Summit for Social Development (WSSD 1995) and The Fifth International Conference on Adult Education (UNESCO 1997) articulated extremely ambitious goals for the reduction of illiteracy, such as the elimination of female illiteracy by the year 2000. While this goal has clearly not been met, meeting the goals of halving by 2025 and again by 2050, as shown in Table 3.4, is certainly a plausible minimal target in a world in which education for all is made a priority.

Finally, there is every reason to believe that aggressive policy actions in the health sector can continue. The goals expressed at such meetings as the international World Summit for Social Development (WSSD 1995) have been quite optimistic, though imprecise. A specific near-term goal is for life expectancy at birth to be greater than 60 years for every country of the world (WHO 1997a). Ultimately, life expectancy in developing regions can be expected to approach that of the OECD regions. A reasonable intermediate goal is that average life expectancy in all countries exceeds 70 years by 2025.

The social goals for the sustainability transition that we have introduced here set a challenging agenda for development. Even as general economic growth tends to drive absolute poverty down, trends in population growth and increasing income disparity tend to drive it higher. Indeed, as will be seen in the next chapter, the *Market Forces* scenario analysis shows that conventional development assumptions alone are not likely to lead to these reductions. But if a policy campaign is launched, it can unleash a virtuous circle that acts across the various dimensions of human deprivation. For example, reducing hunger also reduces vulnerability to disease which, in turn, improves access to livelihoods and food entitlements.

As expressions of different faces of poverty, the problems represented by the various indicators have common roots. In the scenario analysis, it is convenient to focus attention on a single indicator of poverty in order to highlight results and to bound the technical analysis. Specifically, we concentrate on hunger-reduction targets in evaluating the sustainability of *Market Forces* scenarios and in defining targets for *Policy Reform* scenarios. A reduction in hunger is correlated with the alleviation of the entire nexus of unfulfilled basic needs associated with absolute poverty.

Diminishing absolute human deprivation is a necessary condition for social sustainability, but it may not be sufficient. In addition, improved social equity within societies may be an important goal in its own right, not merely as one mechanism for eradicating poverty. At issue is the possible link between social equity and social cohesion – between the perceived sense of fairness in the distribution of wealth and privilege in a society on the one hand, and allegiance to prevailing social institutions on the other. Inegalitarian societies are likely to be susceptible to relatively high rates of political violence and social instability (Gurr 1968, Muller and Seligson 1987, Muller 1988), though the detailed empirical findings are debated (Wang 1993). We must be satisfied here with simply noting the linkage between social sustainability and distributional equity, and observing that strategies for alleviating poverty by moderating social inequality may also enhance social stability.

Environmental objectives

In principle, any significant and irreversible changes to the Earth's environment are not sustainable, since they diminish the opportunities for future generations. In practice, there is an inherent tension between such a strong formulation of the sustainability principle and more immediate goals, such as meeting current human needs and aspirations. Expanding food production may require clearing land for farming or withdrawing more water from streams for irrigation – putting additional pressure on natural ecosystems. Providing the jobs, products, and lifestyles of an industrialized, urban economy will require expanded use of energy, which potentially increases pollution.

The proper balance between the goals of sustainability in the long term and development in the near term is not easily resolved. The relative weight one puts on the rights of current generations, future generations, and the living world has strong ethical and moral dimensions. Complicating the dilemma further is a lack of scientific certainty on many critical issues, such as: how will the global climate be modified in response to rising greenhouse gas concentrations in the atmosphere; how much pressure from human activities can ecosystems withstand before collapsing?

The discussion here will focus on a limited set of environmental objectives. As with the social indicators, international conventions and agreements provide a starting point for selecting environmental criteria for sustainability. There is broad agreement, for example, that the stratospheric ozone layer and the global climate should be protected and that urban and industrial pollution – especially toxic pollution – should be curbed. There is agreement that various ecosystems should be preserved so that the services they provide – from water purification to nutrient recycling to providing habitats for diverse species of plants and animals – can continue, and that future generations can enjoy the natural beauty and genetic wealth of forests, coral reefs, and the Earth's biological bounty. There is agreement that steps need to be taken to maintain the fertility of soils and prevent degradation or desertification.

Beyond such agreements, however, there is as yet no consensus on how to measure progress toward sustainability, although many different groups have proposed sets of indicators for this purpose (Munasinghe and Shearer 1995, Moldan *et al*. 1997, OECD 1998, CGSDI 2001). Moreover, lack of data is a serious barrier to making many proposed indicators operational. The indicators chosen for use in this report constitute a minimum but hardly sufficient set.

However measured, global environmental trends are not encouraging. Despite improvements in reducing some local environmental pollution in OECD countries, pollution levels remain high and are rising rapidly in most developing regions. Pressures on the global commons – the atmosphere and the oceans – are steadily increasing (UNDPCSD 1997). For example, most marine fisheries are now fished at or above sustainable levels (FAO 2000f) and the incidence of toxic phytoplankton blooms in coastal waters is rising (WRI 1994). Pressures on forests, coral reefs, and other vital ecosystems, and on such renewable but finite resources as freshwater, are also escalating (Bryant *et al*. 1997, Burke and Bryant 1998).

A transition to sustainability will require reversing these trends. For the purpose of setting plausible *minimum* targets, we assume that such a transition should be well underway by 2025, crest by 2050 and be completed in the second half of the twenty-first century. This will require beginning to "bend the curve" of the various indicators away from trend projections by 2025 and showing substantial improvements in environmental quality worldwide by 2050. These criteria applied to climate change, for example, would require that global emissions of greenhouse gases peak no later than 2025 and that atmospheric concentrations begin to stabilize by 2050, with stabilization completed by 2100.

Indicators and targets are proposed in Table 3.5 for five types of environmental issues: climate change, resource exploitation, toxic emissions, freshwater problems, and ecosystem pressure. The first have been driven primarily by industrial activities and the demands of

Table 3.5 Global environmental targets

Region	Indicator	1995	2025	2050
Climate				
World	CO_2 concentration	360 ppmv	Stabilize at **<450 ppmv** by 2100	
	Warming		**<2.0°** rise between 1990 and 2100	
	CO_2 emissions		**<700 GtC** cumulative, 1990–2100	
OECD	CO_2 emissions rate	Various and rising	**<65% of 1990**	**<35% of 1990**
Non-OECD	CO_2 emissions rate	Various and rising	Increases slowing, energy efficiency rising	Reach OECD per-capita rates by 2075
Resource use				
OECD	Eco-efficiency	$100 GDP/300 kg	**4-fold increase** ($100 GDP/75 kg)	**10-fold increase** ($100 GDP/30 kg)
	Materials use/capita	80 t	**<60 t**	**<30 t**
Non-OECD	Eco-efficiency	Various but low	Converge toward OECD practices	
	Materials use/capita	Various but low	Converge toward OECD per capita values	
Toxics				
OECD	Releases of persistent organic pollutants and heavy metals	Various but high	**<50% of 1995**	**<10% of 1995**
Non-OECD	Releases of persistent organic pollutants and heavy metals	Various and rising	Increases slowing	Converage to OCED per-capita values
Freshwater				
World	Use-to-resource ratio	Various and rising	Reaches peak values	Decrease if feasible; below 1995 value in high-stress areas
	Population in water stress	1.8 billion (32%)	**<3 billion** (<40%)	**<3.5 billion**, begins decreasing (<40%)
Ecosystem pressure				
World	Deforestation	Various but high	No further deforestation	Net reforestation
	Land degradation	Various but high	No further degradation	Net restoration
	Marine over-fishing	Fish stocks declining	Over-fishing stopped	Healthy fish stocks

modern lifestyles, while water and ecosystem stress are also associated with poverty and growing populations. Many other issues, such as destruction of coral reefs and other fragile habitats, the introduction of invasive exotic species, and additional local water, air, and soil pollution concerns, would need to be included in an exhaustive consideration of sustainability targets. But the five areas we have identified provide a broad and manageable framework for considering alternative scenarios of global development.

To be politically acceptable, equitable sustainability targets at the national level must reflect each country's historic contribution to the problem. This is codified in international environmental treaties, such as the Montreal Protocol on stratospheric ozone and the Kyoto Protocol on climate change, where high-income countries have assumed the obligation of acting first and assisting developing countries as they more gradually align their economies with environmental constraints.

With this in mind, the long-range sustainability targets introduced in Table 3.5 call for substantial decreases in the environmental pressures from OECD economies. At the same time, the targets for developing countries acknowledge that the process of development and industrialization must continue in these regions, with gradual convergence toward the improved OECD environmental standards.

Climate

The long-term goal for climate, as formulated in the Framework Convention on Climate Change, is to stabilize concentrations of greenhouse gases in the atmosphere at safe levels, although there is as yet no agreement on when, and at what concentration levels, stabilization should occur. The impacts of global climate change are complex as alterations in temperature, hydrology, and sea level ripple through human and natural systems. Regional vulnerability would vary to the more frequent and extreme droughts, floods, and weather events that climate change is likely to bring. Societies and ecosystems can mute these impacts through adaptation and migration, but only if change is sufficiently gradual.

In setting targets for climate change, then, a critical question is the rate of change that allows for adaptation. To proceed, we first identify a maximum change in global mean temperature for the protection of ecosystems. Next, we link this increment in temperature to a maximum target for the atmospheric concentrations of greenhouse gas, focussing on carbon dioxide, the major greenhouse gas. Then, we translate this into constraints on global emissions of carbon dioxide emissions. Finally, allowed global emissions are allocated to countries, taking into account historic patterns and future equity considerations.

Change in mean global temperature is a key indicator of the degree of general climate alteration and of the pressure on ecosystems (Krause *et al*. 1989, Rijsberman and Swart 1990, McCarthy *et al*. 2001). As temperature rises, the tendency is for ecosystems to compensate by gradually migrating toward higher latitudes and altitudes. Their viability would be threatened if the movement toward the poles and higher elevations was unable to keep pace with changing climate conditions. If the absolute change of temperature over the next century were too high, vulnerable species and ecosystems would be threatened.

Climate stabilization requires the eventual stabilization of greenhouse gas concentrations in the atmosphere. Atmospheric concentrations are determined by the rate at which the gases are emitted to, and absorbed from, the atmosphere as a result of both human activity and natural processes. Climate models connect assumptions about future emissions of greenhouse gases, atmospheric concentrations and changes in *equilibrium* temperature relative to preindustrial levels. Due to inertia in the climate system, it may take several decades to reach such equilibrium; therefore, a target for *experienced* temperature changes between 1990 and 2100 will correspond to a higher ultimate committed change.

Upper limits have been proposed for both the absolute change in temperature and for the rate of temperature change. Considering the rate of change, some studies suggest that warming should occur no faster than 0.1°C/decade on average between 1990 and 2100 to allow many – but not all – ecosystems to adapt (Rijsberman and Swart 1990, Hare 1997). But the scientific literature on the links between rates of temperature change and capacity for ecosystem adaptation is limited. The International Panel on Climate Change summarizes the connection in a recent report:

> Some plant and animal species ... and natural systems ... could be adversely affected by regional climatic variations that correspond to a less than 1°C mean global warming by 2100. With mean warming of 1–2°C by 2100, some regional changes would be significant enough so that adverse impacts ... would become more severe ... and additional species and systems would begin to be adversely impacted. Warming beyond 2°C would further compound the risks.
>
> (McCarthy *et al.* 2001)

This suggests the minimum goal that temperature should rise less than 2°C between 1990 and 2100, with eventual stabilization under 2°C.

The relationship between temperature change and carbon concentrations is illustrated in Figure 3.3, which shows two trajectories, one leading to an atmospheric CO_2 concentration of 450 parts per million by volume (ppmv) by 2100 and the other of 650 ppmv. In connecting concentrations to impact, an important variable is sulfur emissions. Sulfur compounds are precursors to aerosols, fine particles suspended in the atmosphere that tend to cool the atmosphere by reflecting incoming solar radiation (McCarthy *et al.* 2001). Thus,

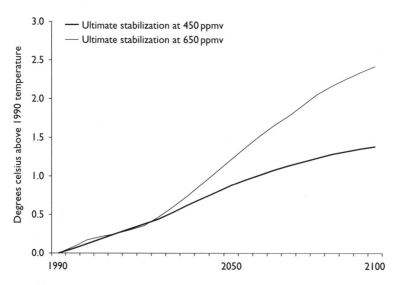

Figure 3.3 Temperature change under different emission trajectories.

Notes: Generated using MAGICC Version 2.4 (Hulme *et al.* 2000).

high sulfur emissions tend to counteract the warming effect of the greenhouse gases. The abatement of sulfur emissions, which contribute to air pollution and acid rain, is itself a desirable goal for sustainability. But, ironically, such reductions would have the indirect consequence of amplifying climate change, all else equal. Figure 3.3 assumes low sulfur emissions consistent with a comprehensive sustainability scenario such as reflected in *Policy Reform* (Chapter 5).

As the figure suggests, limiting global temperature increases to less than 2°C requires that carbon dioxide concentrations stabilize at about 450 ppmv in about 2100. But another important source of uncertainty is the *climate sensitivity parameter* – the equilibrium temperature change associated with a doubling of CO_2 concentration. The climate sensitivity is thought to lie within the range 1.5–4.5°C (Houghton *et al.* 2001). Figure 3.3 assumes a mid-range value, while Figure 3.4 shows global temperature changes for the 450 ppmv trajectory (with low sulfur emissions) for the range of climate sensitivities. This result supports the conclusion that the equilibrium concentration target of about 450 ppmv is likely to maintain climate change within the maximum target for temperature change.

The next step is to convert the 450 target into an acceptable level of carbon emissions. The relationship between atmospheric CO_2 concentrations and cumulative emissions is summarized in Figure 3.5. Shown in the figure are long-term equilibrium concentrations for atmospheric carbon ranging from 350 to 750 ppmv and the corresponding cumulative carbon emissions from human activity between 1990 and 2100 (IPCC 1995a). The carbon sources are primarily from energy combustion and net biomass loss through deforestation and land clearing. The uncertainty ranges reflect differences in the outputs of carbon-cycle models, which simulate the interactions among atmospheric, oceanic, and terrestrial systems. The policy target of 450 ppmv corresponds to a cumulative carbon emissions allowance between 1990 and 2100 in the 640–800 billion tonnes of carbon (GtC) range.

The sustainability target translates into a global budget for cumulative CO_2 emissions from human activities of about 700 GtC. The aggregate global emissions must be allocated

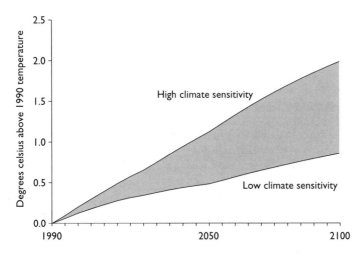

Figure 3.4 Temperature changes for 450 ppmv stabilization concentration.
Notes: Generated using MAGICC Version 2.4 (Hulme *et al.* 2000).

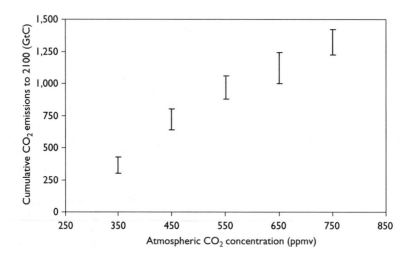

Figure 3.5 CO_2 emissions to 2100 corresponding to different stabilization concentrations.
Source: IPCC (1995a).

to regions and countries. The targets introduced here take into account equity and burden-sharing considerations in the allocation of emissions rights. Relative to 1990 levels, OECD regions are assumed to decrease emissions 35 percent by 2025. All regions approach a common emissions-per-capita target by 2075. Although ambitious, the targets are nonetheless required if climate stabilization at reasonably safe levels is to be achieved in the coming century.

In another sense, the 450 ppmv goal is conservative in that it accepts that considerable global change is inevitable. Also, if the higher values in the climate sensitivity range (i.e., 4.5°) turn out to be more accurate, an additional 0.5° of climate change would occur and the sustainability target would be exceeded. Stronger goals – stabilization at 350 ppmv or even a return to the preindustrial level of 280 ppmv – are arguable from a strong environmental sustainability perspective. Assuming climate sensitivities at the high end of the range, the change of temperature at 550 ppmv would be at levels experienced in the transition to an ice age. It would take 350 ppmv and low climate sensitivities for temperature change to lie within the range of natural fluctuations during the past millennium (Azar and Rodhe 1997).

But given the recalcitrance to date among key countries for adopting vigorous greenhouse gas abatement strategies, it appears that achieving such ecologically driven goals would indeed require a *Great Transition*. Within the *Conventional Worlds* framework the 450 ppmv goal presents a strong, but realizable, challenge for policy.

Resource use

Extracting, refining, manufacturing, transporting, and ultimately disposing of materials is a major cause of pollution and waste in industrial societies. These materials include metals and plastics in automobiles; chemicals that provide the basis for paints, pesticides, and thousands of other products; and minerals, fibers, and other natural resources that comprise everything from clothes to computer chips. As global industrial activity expands several-fold over the next half century, pollution and wastes and the resulting environmental degradation may also expand, unless there is a transition to a cleaner and more efficient – or *eco-efficient* – industrial system.

To simultaneously increase economic output and reduce environmental stress requires profound changes in the technological infrastructure of modern societies. Historically, the throughput of materials into economies has risen with economic growth. In the sustainability transition, material use and economic scale must be de-linked in order to moderate or reduce the pressure on resources and the environment. The process of reducing the throughput per unit of economic activity is sometimes referred to as *dematerialization*. Some dematerialization is already underway. The use of energy and various materials has grown less rapidly than GDP in many industrial countries in recent years. However, the expansion of economic activity has generally outpaced the improvement in efficiency (e.g., increases in miles driven exceed improvements in auto efficiency), so that aggregate use of materials and energy has continued to rise.

In order to stabilize environmental pressure as economies and populations grow, the flow of material into societies must be controlled by designing more durable products, reducing waste streams through reuse of materials, and introducing cleaner production processes. Additional dematerialization would come from a shift in the mix of economic activity itself; for example, from resource-intensive to knowledge-intensive activities. Ultimately, affluent lifestyles based on high material consumption may need to change, as well.

Two useful aggregate indicators of such a resource transformation are the *eco-efficiency ratio*, defined as the ratio of economic output to materials required, and materials use per capita. Current values of the eco-efficiency ratio in the OECD regions are typically about $100 GDP per 300 kg of materials (Adriaanse *et al.* 1997). Material use per capita ranges between about 45 and 80 tons per person per year. Included in the estimates of material requirements are direct inputs of minerals, metals, construction materials, and biomass from either domestic or foreign sources, along with material that is moved or discarded in the process of extraction and processing.

The heterogeneous character of the material stream hampers the formulation of sustainability goals for material efficiency. Eventually, the problem of reducing material intensities will need to be disaggregated into a number of component flows, indicators, and targets in national sustainability plans. Nevertheless, we introduce here provisional aggregate targets to suggest the direction and magnitude of improved eco-efficiency.

We take, as a sustainability target for OECD countries, a tenfold increase in the eco-efficiency ratio by 2050, a widely discussed goal (Factor 10 Club 1995). With this target, $100 of economic output requires only 30 kg of natural resources, as reported in Table 3.5. An ambitious but achievable interim goal is a fourfold increase in the eco-efficiency ratio by 2025. Allowing for economic growth, these targets correspond roughly to a 25 percent reduction in materials use per capita by 2025 and an additional 50 percent decrease by 2050. With this level of improvement in the rich countries, it is possible to prevent global material requirements from increasing, while allowing poor countries to approach an equitable claim on use (Carley and Spapens 1998).

Currently, developing economies have eco-efficiencies even lower than in OECD regions, but with generally much lower values of materials use per capita because the level of output per capita is much lower. In the process of a transition to sustainable development, eco-efficiency in developing regions could improve rapidly if efficient technologies and operating practices of developed countries are adopted and adapted. In some instances, they could leapfrog to advanced technologies. The sustainability target is that developing countries converge toward OECD standards of material efficiency in the course of economic growth.

Toxic substances

Some forms of resource use have such pronounced environmental impacts that more specific objectives are needed. This applies to the widespread industrial use and release of toxic substances. The material and chemical flows underlying a modern industrial economy are immensely complex and are neither well tracked nor well understood. Nearly 100,000 synthetic industrial chemicals contribute to the toxic and hazardous loads on the environment. The pathways into the environment of hazardous chemicals and heavy metals are from mineral extraction and refinement processes, emissions at industrial production facilities, and dissipative losses from materials embodied in products. While some of the substances may not pose environmental or toxicological threats, there is insufficient scientific information for assessing about 90 percent of them.

Heavy metals, persistent organic pollutants (POPs) and other toxic material can persist in the environment and accumulate to dangerous levels in soils, sediments, and the food chain (Jackson and Taylor 1992, Dethlefson *et al.* 1993). Long-lived toxic substances pose a growing threat of uncertain magnitude both to human health and to ecosystems. International efforts on persistent organic substances are focusing initially on limits to the production and use of 12 substances (aldrin, dieldrin, endrin, DDT, chlordane, heptachlor, hexachlorobenzene, mirex, toxaphene, PCBs, dioxins, and furans), but have not yet reached agreement on specific targets or timetables.

The target adopted here assumes that OECD countries reduce toxic substance releases by 50 percent by 2025 with a 90 percent reduction by 2050. Strict application of the precautionary principle, which would avoid environmental hazards even where scientific demonstration of risks is uncertain, would perhaps suggest even stronger protective measures on toxics. The targets here recognize the inertia of formulating policies to change industrial practices and the long turnover time of capital investments, and allow for some unavoidable emissions related to high-priority applications.

Use and emissions of toxic substances in developing countries are far below OECD levels on a per-capita basis, but are rising rapidly and are likely to increase further as industrial activity intensifies. The target set here is that these increases begin to slow by 2025 and converge toward OECD per-capita levels by 2050.

Freshwater

Global freshwater withdrawals have doubled in the past 40 years to nearly 4,000 km^3 per year, far less than the supply of approximately 40,000 km^3 of annual runoff. However, only about 12,000 km^3 of renewable supply is accessible and reliable, since much is lost to floods. Also, in-stream water must be maintained to support such services including diluting pollution and supporting aquatic ecosystems. Freshwater requirements are approaching resource limits (Postel *et al.* 1996).

Aggregate global water accounts mask more severe water stress at the local level. Approximately one-third of the world's people live in areas experiencing some form of water stress. Billions more could join their ranks as population and economic growth drive water demand higher. Water scarcity brings conflict along two fissures; between humanity and nature, and between different human user groups.

Human exploitation of freshwater resources threatens ecosystems and habitats that depend on water in adequate quantity and quality. Free-running rivers are engineered into a vast plumbing system of dams, channels and levees; huge quantities of water are withdrawn from surface and underground sources; and water pollution from point and area sources degrades the quality of water bodies. The ecosystem loss is measured in vanished wetlands, degraded riparian ecosystems, and extinct freshwater species. The reversal of these trends is a cornerstone of the sustainability agenda.

Water scarcity and degradation also fuels conflicts among competing users – between farm and city, between upstream and downstream interests, and between nations sharing common river basins. In the absence of remedial measures, this undermines social cohesion, threatens economic well-being and contributes to international tension in a number of the 260 international river basins (Gleick 2000). Those most tragically affected by inadequate water management are the billion poor people throughout the developing world who lack access to safe drinking water, and the three billion without basic sanitation services. Degraded water supplies are implicated in heightened exposure to pathogenic microbes that cause 250 million cases of waterborne diseases each year.

Until the 1990s, freshwater was the forgotten issue on the sustainability agenda, warranting not a single chapter in the seminal Brundtland report (WCED 1987). Then a series of international assessments brought attention to growing problems of water sufficiency and quality in many water catchment zones (SEI 1997, WWV 2000). In the most severe cases, rivers are running dry, lakes are drying up and groundwater aquifers are being depleted. As thirsty populations and economies grow, the ranks of the water-stressed will swell. The health of people, economies, and ecosystems depends on the prudent management and allocation of finite and vulnerable water resources. Sustainable development requires the provision of sufficient freshwater of good quality to growing economies, while protecting aquatic and marine ecosystems. The multiple roles that water plays in society and nature links water sustainability to the problem of sustainability in general. Water sustainability presents one of the major challenges that human societies will face in the coming decades (SEI 1997, UNDPCSD 1997, UNEP 1997).

The natural hydrological unit for water assessment is the river basin. Nevertheless, the global situation can be illuminated by introducing indicators of water stress at the national level, an exercise that was conducted for the United Nations' Comprehensive Assessment of the Freshwater Resources of the World (Raskin et al. 1997). The basic measure of national water stress that will be used here is the *use-to-resource* ratio. By "use," we mean annual withdrawals, the abstractions from rivers, dams, lakes, and aquifers for use in households, businesses, industry, and agriculture. By "resource," we refer to the average annual renewable freshwater flows. Resource flows include both indigenous rainwater runoff and inflows from international rivers.

As human withdrawals rise in relation to resources, and the use-to-resource ratio increases, competition over water increases. This takes both political and environmental forms, between human groups and between humanity and nature. The character of the competition for water between groups varies with the circumstances in a river basin – agriculture versus other sectors, upstream versus downstream users, and country versus country along shared waterways. The battle between growing aggregate human use and the freshwater needs of ecosystems is ubiquitous, and nature has generally been the loser. Rising demands eventually lead to a condition of water stress in which all claims on freshwater resources cannot be met while preserving water quality and ecological health. Sustainable water development is about moderating stress and thus reducing the potential for conflict within society and between society and nature.

While the degree of water stress and the use-to-resource ratio are clearly correlated, the precise relationship varies across countries. Countries differ in the amount of flow lost through floodwater runoff, in the accessibility of surface and groundwater resources to centers of demand, in down-river claims on water resources, and in the requirements for ecosystem preservation. Nevertheless, when values of the use-to-resource ratio exceed 0.4, a country is generally experiencing a high level of water stress. Indeed, signs of imminent competition between user groups, or between human and natural requirements, can begin when the ratio is as low as 0.1.

In principle, a valid sustainability goal would be that over the next decades pressure on water resources decreases in all areas where water scarcity threatens to impede development or degrade ecosystems. In practice, the momentum of growing water requirements is unstoppable. Water withdrawals have increased by nearly a factor of three since 1950 (Shiklomanov 1997), and the continuing momentum of expanding populations, production and food needs will imply continuing growth in water requirements. As other resources such as petroleum or rare minerals grow scarce, substitutes can be found, but freshwater substitutes do not exist and water transfers between basins are often both expensive and environmentally problematic. If climate change significantly alters hydrological patterns, then water strategies will face a further source of change, the impact of which will be difficult to anticipate. Sustainable water strategies must be flexible if they are to respond to these changes.

In light of these difficulties, a realistic sustainability goal is that, in river basins where freshwater has become, or will become, scarce, withdrawal requirements should peak by 2025. Specifically, in countries where there are indications of freshwater stress (use-to-resource ratios greater than 0.1), the target is for withdrawals to stop increasing by 2025. Since this goal allows for increasing use until 2025, water stress will increase. Indeed, assuming mid-range population growth, the target for 2025 implies that up to three billion people would be subject to water stress conditions. The goal for 2050 is that use-to-resource ratios decline, and that they will not exceed base-year levels in highly stressed areas.

The freshwater targets are rather weak in reference to environmental sustainability criteria alone, since they recognize the inevitability of continuing water stress in many regions. As we shall see in Chapter 5, meeting even these targets is not easy under *Conventional Worlds* conditions. To sharply reduce water stress would require the lower populations, reduced consumption and revised values assumed in a *Great Transition*. In any scenario for sustainability, freshwater sustainability requires a major policy focus aimed at increasing water-use efficiency, reducing losses and enhancing dependable resources.

Ecosystem pressure

Increasing demands for food, fiber, and timber are pressing against the limits of natural ecosystems. Expanding infrastructure to meet the housing, commercial, and transportation needs of growing populations and economies compounds the problem. To avoid serious damage, fragile ecosystems need to be protected, deforestation and destructive logging practices halted, arid lands carefully managed to avoid desertification, tilled lands managed to avoid erosion or other forms of degradation, and the expansion of built areas moderated. The human destruction of the world's natural heritage certainly impoverishes the world's beauty and species diversity. But it is also myopic, since ecosystems are the source of numerous goods and services that underpin healthy economies.

Three indicators serve as imperfect proxies for the state of ecosystems: deforestation, land degradation, and the extent of overfishing in the major marine fisheries. In a future committed to sustainability, such as envisioned in the *Policy Reform* scenario, net loss of forested land should be much lower than historical rates (or those in the *Market Forces* scenario). The target is for deforestation rates (the net forests lost per year) to reach zero before 2025 in all regions. By 2050, forest cover (and quality) should be increasing. This implies the maintenance of critical levels of ecologically rich natural forests. A growing proportion of the world's expanding need for wood, paper, and other forest products should come from plantations or other sustainably managed forests. Both productive forests and plantations should increasingly be managed so as to enhance their ecological quality – their capacity to sustain diverse biota and deliver a range of ecosystem services. Similarly, land degradation rates (e.g., the land lost to agriculture per year as a result of chemical or physical erosion) should also slow to zero by 2025. Finally, overfishing should be curtailed so that the world's fish stocks can rebuild themselves to healthy levels.

Moving toward a sustainable world

This array of social and environmental global indicators and targets provides an operational framework for exploring the quantitative requirements for a sustainability transition and for devising strategies for inducing such a transition. The environmental criteria for sustainability proposed here are ambitious, particularly with a *Conventional Worlds* context. In the spirit of *adaptive management*, sustainability criteria will need to be revised from time to time in response to new information and priorities. Moreover, with scientific understanding of the Earth system still fragmentary, additional environmental concerns will no doubt arise in the years ahead. Nevertheless, our provisional suite of sustainability criteria offers reasonable initial directions for navigating toward sustainable development – and for weighing how far off course current trends may be taking us, the topic to which we now turn.

Chapter 4

Market-driven globalization

From the turbulent and uncertain conditions of the evolving world system, the global development trajectory can branch into contrasting futures. In the scenario framework introduced in Chapter 2, the spectrum of long-range possibilities was organized into three archetypal categories – *Conventional Worlds*, *Barbarization*, and *Great Transitions*. Each of these broad and structurally distinct scenario groups encompasses a range of plausible scenario variations. To keep the taxonomy manageable, we have introduced two variants for each category.

We begin our exploration of the scenarios with *Conventional Worlds*, a vision of the future in which the dominant drivers of change persist; economic globalization, the information revolution, and cultural interchange advance global interconnectedness; and developing and transitional countries gradually assimilate into the liberal market development model. In this chapter we focus on the *Market Forces* variant, where free-market ideology and the economic dimension dominate global development. In the next, we consider the *Policy Reform* variant, where government-led initiatives seek to guide and control the global market to ensure its compatibility with sustainable development as expressed in an array of environmental and social goals.

Market forces scenarios

Market Forces scenarios illustrate the tension between the current patterns of world development and sustainability goals (Raskin *et al.* 1998). This is a story of the future in which demographic, economic, environmental, and technological trends unfold without major surprise. Continuity, globalization, and convergence are key characteristics of world development – institutions gradually adjust without major ruptures, international economic integration proceeds apace, and the socioeconomic patterns of poor regions converge slowly toward the model of the rich regions. The contradictions of forging a global market system – social polarization, violent resistance, and environmental degradation – are contained.

Market Forces scenarios are distinguished by the absence of strong, coordinated and proactive policies for achieving sustainability goals. Rather, policy focuses heavily on the removal of barriers to markets at national and international levels (OECD 1997) and modernization of developing and transitional economies. In the international arena, an integrated world economy is promoted through liberalization of trade and finance. At the national level, the push is for thorough and rapid structural adjustment, deregulation, and privatization.

The trends and forces already in the pipeline drive *Market Forces* scenarios forward. The global population becomes larger, more urban, and older. Expanding global trade, financial transactions, and capital flows drive economic growth. The free cross-border flow of goods, services, and capital becomes a reality. Transnational corporations increasingly dominate economic activity and enjoy growing political influence. Spurred by modern media and information technology, people everywhere, and especially the young, are drawn to consumerism. Materialism and individualism spread as core human values.

Despite economic growth, extreme income disparity between rich and poor countries, and between the rich and poor within countries, remains a critical social trend. Environmental transformation and degradation is a progressively more significant factor in global affairs. Technology is a continuing source of change through advances in information technology, biotechnology, and myriad innovations that change the way we produce and consume. The trends are allowed to play out without major changes in policy.

An illustration

The *Market Forces* narrative is compatible with a range of specific outcomes. Nevertheless, a quantitative sketch of the broad contours of the scenario illuminates the broad implications of the scenarios. The illustration incorporates demographic, economic, and technological projections based on mid-range, or "business-as-usual," projections by international studies that share the basic premises of the scenario.

A global overview of the scenario is presented in Figure 4.1. Detailed results by region are summarized in the Annex, while the basis for the figures is the subject of this chapter. By 2050, world population increases by more than 50 percent relative to 1995, average global income (expressed as GDP per capita) grows to over 2.5 times the 1995 value, and economic output more than quadruples. These increases in population and economic output set the overall scale for human activity, modified by changes in consumption and resource-use patterns.

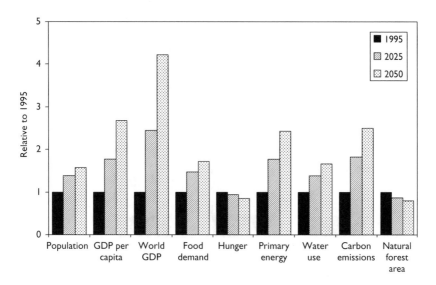

Figure 4.1 Global overview of the *Market Forces* scenario.

Food requirements grow faster than population, increasing by over 70 percent between 1995 and 2050, as income growth leads to higher food consumption and changes in diet structure. Nevertheless, the world's undernourished population declines only slightly, although the number of hungry people as a fraction of total global population decreases significantly. The hunger-reducing effects of rising incomes are largely offset by increases in population and inequality in the distribution of income.

Resource requirements grow less rapidly than the economy as a whole in the scenario. This is a result of two effects; improving efficiency of use, and a gradual shift in the composition of economy toward less resource-intensive activities. Requirements for energy and water increase by factors of 2.4 and 1.7, respectively, compared to an increase in economic output by a factor of 4.2. The carbon intensity of energy production remains essentially unchanged over the course of the scenario, as global carbon dioxide emissions from energy use, a major driver of global warming, increase by a factor of 2.5 over the scenario period, slightly higher than the increase in energy use. The deployment of renewable and nuclear energy sources increases everywhere, but fossil fuel use remains dominant. Increasing land requirements for agriculture and human settlements lead to a 20 percent drop in forest area.

Growing populations and increasing urbanization

Despite slowing population growth, the mid-range United Nations population projections show a substantial increase of 3.2 billion people between 1995 and 2050. Global population in the scenario is presented in Figure 4.2, and regional breakdowns are shown in Annex D-1. The OECD and transitional regions have largely passed through their demographic transitions, so nearly all the world population growth occurs in the developing regions. Of the net increase of 3.2 billion people by the year 2050, only 3 percent is in the OECD regions, while the populations in Eastern Europe and the FSU actually decrease slightly in the scenario.

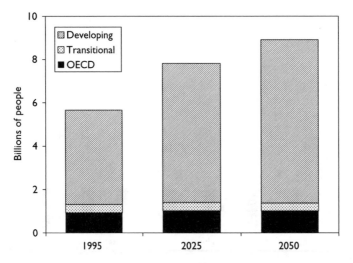

Figure 4.2 UN mid-range population projections to 2050.
Source: UN (1999).

The *Market Forces* scenario incorporates mid-range projections for urbanization (see Annex D-2). Urban population in developing regions more than doubles between 1995 and 2025, and more than triples between 1995 and 2050. Globally, the urban population rises from 2.6 billion in 1995 to 6.6 billion in 2050, as the proportion of the world's population living in cities rises from roughly one-half to nearly three-quarters. These trends pose immense challenges for urban infrastructure investment, employment generation and institutions of governance.

Expanding economies and gradual convergence

Basic premises of the *Market Forces* development include increasing globalization and inter-dependence, economic growth and progressive inclusion of developing countries into the nexus of global markets. Economic growth rates are higher in poorer than in richer regions as economies steadily converge. The economic growth rates assumed in the illustrative scenario are compatible with mid-range assumptions of various global scenarios, such as those of the International Panel on Climate Change, which are consistent with these expectations (IPCC 1992, Houghton *et al.* 2001). Regional trends in GDP and income are shown in Annexes E-1 and E-3.

As noted in Chapter 3, international comparisons of economic output are converted to a common currency using PPP, rather than the standard MER. By comparing local costs of a common set of goods and services, the PPP method provides a more precise estimate of the relative magnitudes of economic activity. We use the notation "GDP_{PPP}" and "GDP_{MER}" to distinguish GDP estimates expressed in PPP and MER terms, respectively. In developing countries, GDP_{PPP} is typically greater than GDP_{MER}, often significantly so. In the high-income OECD countries, they tend to be about the same. Comparable items are typically less expensive in developing countries (in dollars, using MER) than, say, in the United States. In the scenarios, therefore, as countries develop, the two expressions for GDP tend to converge, implying that the growth rate of GDP_{PPP} will be less than of GDP_{MER}. In comparing GDP projections using the two methods, we assume that the ratio of GDP_{PPP} to GDP_{MER} approaches a value of one as income approaches one-half the US income, based on an examination of the historical data shown in Figure 4.3.

Global GDP_{PPP} increases from \$35 trillion in 1995 to over \$145 trillion in 2050, an average growth rate of 2.7 percent per year. The higher growth rate in developing region economies, about 3.3 per year, causes their share of world product to rise from about 40 percent to close to 60 percent. The immense scale of the economy implied by these assumptions is illustrated by the case of China+, where the GDP_{PPP} in 2050 is greater than the whole of the OECD today.

In all regions, GDP_{PPP} growth exceeds population growth, so average income (represented by GDP_{PPP} per capita) increases everywhere. Regional trends are shown on Annex E-3. The projections show a gradual convergence of developing regions toward industrialized ones in the sense that the ratio of average income of the OECD regions to that of the non-OECD countries decreases from 6.7 to 5.6 over the next 50 years. At the same time, the *absolute* difference between OECD and developing regions increases from about \$17,800 per capita in 1995 to \$48,700 per capita by 2050.

In general, the scenario is one of wide prosperity by mid-century, in which developing regions approach standards of living enjoyed in Western Europe *c.* 1980, while incomes in OECD regions soar to unprecedented levels. In North America, average incomes reach \$73,000 per capita by 2050, an income accessible only to the top-earning 20 percent of the

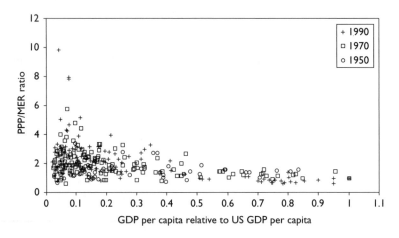

Figure 4.3 PPP-to-MER ratio.

Source: Heston *et al.* (1995).

US population in 1995 (USBC 2000). However, in Africa, GDP growth between 1995 and 2025 is only slightly higher than growth of the urban population, suggesting that the extension of satisfactory infrastructure to mushrooming urban centers will continue to be a big challenge.

As the scale of economic output grows in the scenario, the structure of economic activities changes. The share of GDP from agriculture declines, continuing a pattern long associated with the process of industrialization and development. Moreover, as consumption patterns shift toward greater emphasis on services and light manufacturing, the share of industry in economic output declines. These patterns are reflected in Annex E-2.

Within the industrial sector, the composition of industrial production also changes. The contribution from heavy industry decreases in favor of less resource-intensive sectors, reflecting dematerialization trends in advanced economies (Williams *et al.* 1987, Bernardini and Galli 1993). This shift contributes to declines in average industrial energy, water and material energy intensities, and the muting of environmental impacts per unit of production. But growth in total production exceeds the rate of decline in intensities, so that aggregate environmental impacts continue to rise.

Increasing energy demand

The compatibility of an energy-hungry *Market Forces* scenario and sustainable development is a critical issue, since energy production and use are directly linked to climate change, acid rain, local air pollution, and many other environmental problems.

Today's energy patterns and problems were forged in a long process of industrialization, cultural transformation, and geopolitical change. Since the mid-nineteenth century, global energy consumption has increased by a factor of almost 20. Modern fuel forms – fossil fuels, nuclear power, and electricity – have progressively replaced traditional forms of fuel, such as wood and charcoal.

The *Market Forces* scenario sees continuation of these trends on a global scale. Global primary energy requirements in 1995 were about 390 EJ (see page 198 for a definition of units), with nearly half used by the 16 percent of the world's population that lives in the OECD regions (Annex En-1). The energy mix is dominated by fossil fuels – coal, petroleum, and natural gas – which contributed 80 percent of the total. Nearly 30 percent of primary energy requirements is lost in conversion to "final fuels" such as electricity and refined petroleum products. These final fuels are what are ultimately consumed by end users in households and businesses, and for transportation.

Global patterns for final fuel use are shown in Figure 4.4. Breakdowns are presented by type of fuel, economic sector, and geographic region. In the scenario, changes in sectoral activity emerge from the demographic and economic assumptions outlined above, while energy technology and fuel mix trends are compatible with typical "business-as-usual," mid-range global energy assessments (IPCC 1992, Raskin and Margolis 1995, WEC/IIASA 1995).

In the analysis, the structure of sectoral demand has been decomposed into numerous subsectoral activities such as major types of industries, transportation modes, and household end-uses. The structure of demand and the intensity of energy use and choice of fuel are all expected to change over the next several decades. When the various factors are compiled, global aggregate energy intensity (the ratio of energy requirements to GDP) decreases from 11 to 6 MJ per dollar GDP$_{PPP}$ between 1995 and 2050 (Annex En-1). While fossil fuels continue to dominate in energy production (see Annex En-2), there is a notable absolute expansion in renewable and nuclear energy, by factors of 1.9 and 2.4, respectively.

Note that, in comparing the shares of different fuel forms, the contribution of renewable energy that is used for power production must be converted to primary energy equivalents for comparison to fossil fuels. The convention used here is that hydropower, wind, and solar are reported in their electrical energy equivalent assuming 100 percent conversion efficiency, following the International Energy Agency (1999). Other studies express renewable energy contribution by estimating the amount of fossil fuel energy that would be required

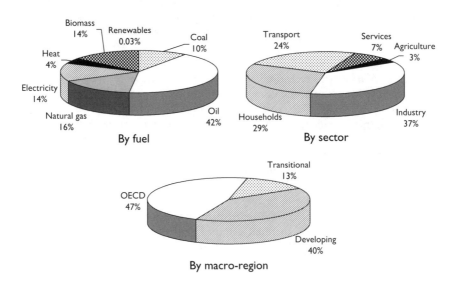

Figure 4.4 Global final energy consumption shares in 1995.

to generate the equivalent amount of electricity (WEC/IIASA 1995). Since the latter method increases reported values of renewables by about a factor of 3, care is needed in comparing results.

In developing regions, a sixfold increase in GDP between 1995 and 2050 converts to an increase of energy by a factor of 3.8 in the scenario. Developing-region demand reaches 60 percent of world demand by 2050, compared to 40 percent today. By 2050, the energy requirements of China+ and South Asia are comparable to those of North America. *Market Forces* would vastly expand the volume of international trade in energy resources and shift the regional patterns of demand. As we shall see, the "great game" of global oil politics would reach a more complex and problematic phase.

Changing structure of food demand and production

The growth of world food requirements in the scenario depends on several factors. Global population drives the overall scale of demand, increasing by nearly 60 percent between 1995 and 2050. In addition, as incomes rise so do food demands per capita (see Annex F-1). Combining these effects leads to a 70 percent increase in food demand over the course of the scenario. In addition, the composition of diets shifts toward more meat consumption as affluence spreads. In developing countries the contribution of animal products to diets rises from 11 percent to 15 percent and the total demand nearly triples between 1995 and 2050.

These scenario patterns would not be unprecedented. Between 1961 and 1995, population in developing countries increased by 60 percent, food demand doubled, and animal products consumption rose by a factor of 3.5. At the same time, hunger was rampant during this period, not due to an absolute food shortage, but rather to the lack of access among the poor (Sen 1981). Despite the growth in aggregate demand, agricultural land in developing countries grew by only 16 percent between 1961 and 1995 (FAO 1999b). Rather than the extensive expansion of agriculture to meet growing food demands, production increases were met largely through intensive increases in yields. The Green Revolution was important in increasing output per hectare, through the introduction of new crop varieties and high inputs of irrigated water and chemicals, and by increasing the number of harvests per year.

Yield-increasing innovations in agricultural practices and technology are expected to continue in a *Market Forces* scenario. In addition to conventional techniques for improving the performance of cultivars, biotechnology has the potential for increasing yields, while reducing chemical inputs, conserving water, and improving the nutritional content of food products. But biotechnology also risks reducing crop diversity, releasing pest-resistant organisms into fragile ecosystems and increasing the dependency of farmers on transnational agribusinesses. The rapid introduction of genetically modified species in the 1990s was insensitive to consumer concerns and environmental uncertainties. However, if it proceeds with greater prudence and the necessary precautions are in place, biotechnology will have an important role to play in feeding the world in the scenario.

On the other hand, it cannot be assumed that past rates of yield increase can be carried forward indefinitely. Already, there are signs of slowing yield growth in some regions (Brown *et al.* 1997, Conway 1997, Pinstrup-Anderson *et al.* 1997). In the developing Asian countries, among the main beneficiaries of the Green Revolution technologies, the average rate of increase of cereal yields from 1991 to 1998, while higher than the average increase over the decade of the 1960s, was below the average rate of the 1970s, and nearly

20 percent below the rate for the 1980s. At the same time, the high levels of chemical and water inputs, wastes from confined livestock operations, salinization on poorly drained irrigated land, and other factors, are creating increasingly severe environmental impacts.

The agricultural scenario starts with human dietary and industrial demands for agricultural products and translates them into requirements for land, water, and nutrient inputs as shown in Figure 4.5 (Leach 1995, Kemp-Benedict *et al.* 2002). Both crops and animal products are consumed for food and feed, while some agricultural products are used as fuel or as industrial feedstocks. Crop production can be irrigated or rainfed, while livestock can be grazed, fed on grain or fodder, or fed crop residues, wastes and other informal sources.

On the demand side, *Market Forces* scenarios assume continuity with past changes in food consumption patterns as incomes rise. Diets in all regions tend toward Western European patterns over the course of the scenario, in both average calorie intake and in the proportion of calories derived from animal products (Annex F-1). Requirements for food and other agricultural products rise with income while, at the same time, higher-valued foods are slowly substituted for cereal staples as incomes increase. In North America, where health concerns continue to cause a gradual decline in reliance on animal products, the share of animal products in diets shifts toward the lower Western European level.

On the production side, agricultural output increases in all regions. Requirements and supply of meat and milk products are shown in Annex F-2, of fish and seafood in Annex F-3 and of crops in Annex F-4. In the OECD regions as a whole, production of animal products increases by over 20 percent, and crop products increase by 40 percent, between 1995 and 2050. In Eastern Europe and the FSU, production rebounds from the collapse during the 1990s, and then expands as these regions become significant agricultural exporters. Rapidly rising food demands and limited scope for expanded production in South Asia converts the region from a net crop exporter in 1995 to an importer of over 15 percent of requirements by 2050. Crop imports as a fraction of total requirements increase in the Middle East from about 40 percent in 1995 to almost 45 percent by 2050, and in Africa from 19 to 28 percent. China+ maintains near self-sufficiency in rice production, while Southeast Asia

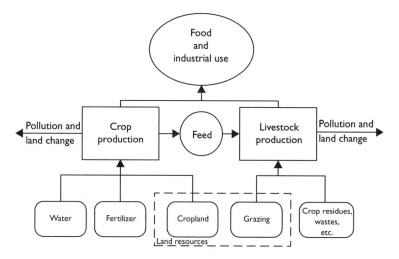

Figure 4.5 The agricultural production system.

becomes a significant exporter, even as rice declines as a fraction of the total diet in these and the other Asian regions in preference to other grains, meat and higher-valued crops.

Underlying the increase in crop production in the scenario are assumptions on yield improvements, and expansion of cropland. The scenario is broadly compatible with the results of recent assessments of the prospects for world agriculture over the next two to three decades (Alexandratos 1995, Rosegrant *et al.* 1995, FAO 2000a). It shares their optimism of sustained improvement in agricultural performance. Figure 4.6 shows average yields historically and in the scenario. Yields gradually increase in all regions (Annex F-5), the result of greater reliance on irrigated farming (Annex F-7), and assumed improvements in plant varieties, farm practices, and biotechnology applications. Global average cereal yields increase by about 1.3 percent per year between 1995 and 2025, and 0.9 percent per year between 1995 and 2050.

While these increases are substantial, they are less than the 2.2 percent cereal yield improvement per year between 1961 and 1995, in part due to constraints on the expansion of irrigated farming (Leach 1995). In addition to the intensive increases in yields, farmland expands extensively into forests and pasture lands (Annexes F-6, P-7, and P-8). The expansion is checked by the reduced availability of suitable land, limited freshwater resources, and competition for land. Globally, farmland increases some 10 percent, mostly in Africa. Irrigated farm area increases slowly from 256 to 309 Mha, remaining just under 20 percent of total farmland throughout the scenario (Annex F-7). In addition, the *cropping intensity*, or average number of harvests per year, gradually grows, further increasing farm output per area (Annex F-5).

The nearly doubling of meat production in the scenario (Annex F-2) drives expansion in both grazed livestock farming and feedlot production, which relies on crop products to

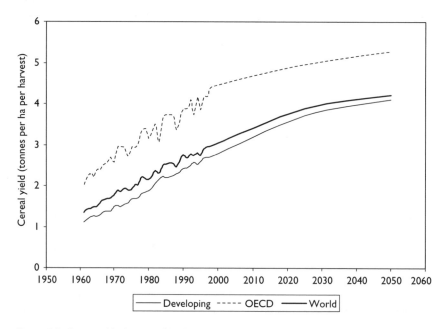

Figure 4.6 Crop yields, historical and scenario.

Source: Historical data from FAO (1999b).

nourish animals. The use of informal feed sources, such as crop residues and household wastes, declines as commercial livestock production systems spread. Grazing land area increases globally at an average of 6.5 Mha per year between 1995 and 2050, compared to an average growth in cropland of 3.0 Mha per year (Annexes P-7 and P-8). Pasture and grazing land expands into forests, ungrazed grasslands, and marginal land as the scenario unfolds.

The expansion in grazing land would be even greater in the scenario were it not for two countervailing processes. First, the productivity of grazing – the output of livestock products per unit of land – improves. The improvements are sharpest in developing regions due to better quality herds, reduction of animal diseases and improvements in pasture management. The area of land required to produce a tonne of beef drops by 15 percent over the 55 years of the scenario. While significant, this improvement is less than an estimated 40 percent increase in productivity in Latin America between 1961 and 1994. In the scenario, livestock diseases are controlled. Today, animal trypanosomiasis, for example, can limit cattle densities in Africa to 30 percent of their potential (Swallow 1999), while insufficient veterinary and disease-control services can favor the adoption of hardy, but lower-yielding animals (Homewood and Rogers 1991).

Second, the trend toward increasing reliance on feedlot production continues in the scenario. In 1995 an estimated 21 percent of global requirements for animal feed was from crop products, while by 2050 this rises to 31 percent, with far more rapid increases in developing regions (Annex F-2). Of course, the increased requirements for cropland are far less than the grazing land spared. Another factor increasing productivity is that incremental growth in global demand tends to be met by increased output from countries with higher-than-average productivity. The changing regional mix of production in the scenario contributes an increase of 20 percent in average global productivity.

Fish and seafood demands are met either from wild fisheries or from fish farming. Wild fisheries are already highly utilized, with many natural fisheries already exploited at unsustainable levels. Aquaculture production has expanded to meet growing demands, a pattern that continues in the scenario. But, as the most economically attractive and environmentally acceptable sites come into production, the incremental costs of further expansion rise. These production constraints imply higher consumer prices, which moderates demand growth. In the scenario, fish and seafood consumption increases from roughly 130 Mt per year currently to 170 Mt by 2050. Regional detail is shown in Annex F-3. Assuming that the contribution from overtaxed marine resources remains steady at about 88 Mt, a fourfold expansion of aquaculture is required, as illustrated in Annex P-10.

In this picture, the world produces enough food to supply the growing demand. However, this is a conclusion subject to important caveats: sufficient water must be available to supply the expanded irrigation requirements, serious land degradation and agrochemical pollution must be mitigated, and trade flows must not face impediments. Finally, despite substantial increases in food production, the problem of access to food among the very poor remains, and hunger persists, as we discuss below.

Rising demand for water

Water demand increased sixfold between 1900 and 1995, driven by irrigation expansion, population growth, and economic development (SEI 1997). This level of growth cannot continue indefinitely, since withdrawals are already nearly half of available supplies (Postel *et al.* 1996). Freshwater flows may seem abundant, but much is either lost as floodwater, is too remote for use or is needed in rivers to dilute pollution, maintain aquatic ecosystems and provide such in-stream services as recreation and hydropower generation. As water requirements rise, so does competition over the entitlement to finite water resources. The competition takes multiple forms, depending on the conditions at the river basin level – between the farm and the city, between environmental and human use, and between upstream and downstream riverine countries.

In the *Market Forces* scenario, water requirements for development continue to increase, driven by growing populations, economies, and agricultural requirements. Global water requirements grow by a factor of 1.7 over the scenario period, and in developing regions by factors ranging from 1.7 in China+ to 3.0 in Southeast Asia (Annex P-1). While substantial, the pace of water demand growth of less than 1 percent per year in the scenario is much slower than the historic rate of about 2 percent per year. This moderation relative to the historic pattern is due to several factors: decreasing population growth, slower expansion of irrigated lands, shifts in the composition of economic activity toward less water-intensive sectors, and greater efficiency in water use. The changing sectoral pattern of water use is summarized in Annex P-2. Note that 67 percent of current global freshwater withdrawals is for irrigation, decreasing to 61 percent by 2050. Nevertheless, the absolute requirements for irrigation water increase by close to 50 percent as agricultural land under irrigation expands by 20 percent (Annex F-7), and more water is taken up by higher-yielding plants (Raskin *et al.* 1997).

Environmental risk

Market Forces scenarios reflect a vision of increasing affluence, a transition from agrarian toward industrial economies in the developing world, and increasing consumption of food, services, and material goods. There is much that is positive in this vision, as most people raise their living standards and enjoy more options in the conduct of their lives. The scenario assumes that economic globalization is enabled by a relatively stable global political climate, the modernization of market-supporting institutions and the ascendance of liberal democracy in much of the world, along with greater personal liberty and freedom.

But the positive elements are marred and potentially threatened by negative aspects of the scenario. With international and national policy focused on economic growth promotion, action to mitigate long-term environmental damage is weak. Pressure mounts on undervalued land, water, energy, and mineral resources. With the future deeply discounted in the calculus of economic investment, the global natural heritage of ecosystems, species, and places of beauty erodes. The sustainability goals introduced in Chapter 3 provide a benchmark for gauging the degree and character of environmental unsustainability in the scenario; that is, in the absence of major course changes in technology, demographics, and consumption patterns.

Climate change

The proposed sustainability goal for climate change is to allow most, if not all, ecosystems to adapt to temperature changes. This implies limiting the increase in average global temperatures to no more than 2°C between 1990 and 2100. The *Market Forces* scenario does not meet the target. Annual carbon emissions are plotted in Figure 4.7. Cumulative carbon emissions between 1990 and 2100 would be about 1500 GtC (which includes about

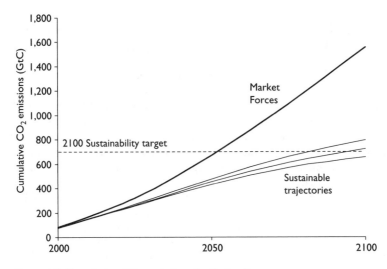

Figure 4.7 Cumulative CO_2 emissions in *Market Forces*.

80 GtC from tropical deforestation). This is about double the target range required to stabilize concentrations compatible with the sustainability goals.

Ironically, the climate impacts of these higher emissions are moderated by the high emissions of sulfur compounds in the scenario. While sulfur dioxide is a noxious contributor to local air pollution, it also is a precursor to atmospheric aerosols that reflect radiation and cool the climate. Nevertheless, the mid-range temperature change in the scenario exceeds the sustainability target. Moreover, neither carbon emissions nor temperature begin to stabilize by 2100. These patterns violate the sustainability targets of limiting emissions to below 2°C between 1995 and 2100, and eventually stabilizing at no more than 2°C.

Global patterns of carbon emissions in the *Market Forces* scenario are reported by region on Annex P-4. Annual emissions more than double over the 1995–2050 period. The share of emissions from OECD regions drops from about 50 percent to 30 percent over this time frame, by a factor of 4.5 between 1995 and 2050, as developing region emissions rise rapidly, driven by population growth and the long-range economic and technological convergence assumptions in the scenario. The changing regional composition of emissions is striking. In 1995, North America is responsible for 1.6 GtC emissions, accounting for 26 percent of the global total. During the first decades of the next century, China+ surpasses North American emissions. Emissions in Africa increase by over a factor of 7 in the scenario.

Despite the rapid rise in developing country emissions in the *Market Forces* scenario, substantial inequities remain. Emissions per capita grow much faster in developing regions, so there is some convergence – the ratio of the value in North America to that in Africa is 22 in 1995 but falls to 9 by 2050 (Annex P-4). Nevertheless, large absolute differences in emission loads persist throughout the scenario period, another expression of the slow process toward the amelioration of current disparities in economic development in the *Market Forces* scenario. Reduction in regional differences in emission rates is an element of the normative goals defining the *Policy Reform* (Chapter 5).

Fossil fuel sufficiency

The question of the adequacy of fossil fuel resources over the long term is closely related to the issues of climate change and sustainable energy. Following the oil crises of the 1970s, concern over petroleum shortages vanished from newspapers and dropped off the policy agenda. Prices dropped, new oil fields have been brought into production and advanced techniques have increased the recovery rates at existing fields.

The complacency over the last two decades about the abundance and reliability of global oil will not persist in a *Market Forces* future, and already has resurfaced as dependency on oil from the politically unstable Middle East continues to grow. The rapid growth in demand in the scenario would put heavy pressure on dwindling global petroleum resources. The dependence of OECD countries on imported Middle East oil, and their vulnerability to export cutbacks, would need to be addressed to ensure the global economic cooperation that underlies the continuity of the scenario. Solving the problem of reliability of global oil markets is critical to the economic and political sustainability of *Market Forces*.

Estimates of ultimately recoverable reserves – the total of previously extracted, proved and undiscovered reserves recoverable – are of course uncertain. By the late 1990s, approximately 800 billion barrels (Gbbl) had been extracted and 800 Gbbl were recorded as proved reserves. The debate concerns the additional reserves that are yet to be discovered, a figure that is difficult to estimate since it depends on a number of geological and economic assumptions. Some analysts argue for an estimate of ultimate conventional petroleum reserves (oil and natural gas liquids) as low as 1,800 Gbbl, leaving only 200 Gbbl to be discovered and 1,000 Gbbl to be extracted (Campbell 2001). At the other extreme, a recent study assumes nearly 2,100 Gbbl of remain, but these are based on "P05 percent estimates," that is, reserves that are only 5 percent likely to be met or exceeded.

For the scenario, ultimate reserves are taken to be about 3,200 Gbbl (WEC 1995, USGS 2000). This implies that estimated undiscovered petroleum reserves were approximately 1,600 Gbbl in 1995. With demand at about 26 Gbbl per year in 1995, and growing at an average annual rate of 1.7 percent (Annex En-2), these reserves would be depleted around 2035 (Annex En-5). According to standard theories of oil exploitation, production decreases may occur much sooner, perhaps as early as 2010 when half of the global resource will have been extracted (Hubbert 1956, Mackenzie 2000, Deffeyes 2001), or by 2003 if low-end estimates of ultimate reserves are assumed. The geopolitics of oil becomes a resurgent theme in the scenario as industrialized regions increasingly depend on imports from the Middle East and Latin America (Swart 1996).

Of course, the gap between oil supply and demand will be closed in practice. Supply might be increased by vast new discoveries of conventional sources, though this is not likely, or by unconventional options such as tar sands and oil shale. However, exploitation of unconventional resources is likely to pose severe environmental problems and to be much more costly than conventional crude. On the demand side, the pressure on oil resources could be curtailed through massive fuel switching and efficiency improvements, options taken up in the *Policy Reform* scenario (Chapter 5).

The geographical distribution of remaining fossil fuel resources has important geopolitical implications. Total proved and additional reserves are heavily concentrated in the Middle East (43 percent) and Latin America (25 percent). Considering proved reserves only, the Middle East's share is 65 percent. The ubiquitous rise of oil demands in the scenario and unequal regional endowments of petroleum resources lead to increasing oil trade and dependency on imported oil. By the year 2025, North America, Europe, OECD Pacific, and the FSU are almost entirely reliant on imports, and Africa meets half of its requirements through imports. Global oil balances are dominated by the Middle East and Latin America, with each accounting for about 40 percent of total production.

These patterns have profound ramifications for international politics, the global economy, and international security (Raskin and Margolis 1998). The oil shocks of the 1970s suggest the potential vulnerability of the world economy to oil price manipulations. The United States and its allies fought the Gulf War in part to ensure access to Middle East oil. That conflict was a visible manifestation of the politics of oil, which plays out more quietly in foreign policy decisions, such as support for undemocratic and unpopular regimes in the interests of oil security.

As the link between global economic stability and petroleum trade security strengthens in the scenario, the world system becomes increasingly vulnerable to highly disruptive terrorist attacks, such as blocking critical shipping lanes by sinking huge oil tankers and/or destroying large production facilities. The *Market Forces* path would progressively intensify the role of the politics of oil in world affairs, with the attendant risks of economic vulnerability, heightened international tensions, and war. This would constantly challenge the continuity and growth assumptions of this development path.

The cumulative extractions of natural gas stood at about 50 Tm^3 in 1995. Remaining global proved and undiscovered natural gas reserves are estimated at about 250 Tm^3 as of 1995 (WEC 1992, 1995). In the *Market Forces* scenario, cumulative requirements for natural gas between 1995 and 2050 are about 200 Tm^3, implying growing stress on reserves, though not as severe as petroleum (Annex En-5). Natural gas reserves are heavily concentrated in the FSU (35 percent) and the Middle East (32 percent).

International natural gas trade is more constrained than for oil. Most current trade occurs between contiguous regions using pipelines because of the technical complexity and high costs of intercontinental transport. Substantial quantities are piped from the FSU to Europe and from Canada to the United States, while liquefied natural gas (LNG) is shipped from South and East Asia to Japan, and from Northern Africa to Western Europe. Although now relatively expensive due to high processing and shipping costs, LNG becomes progressively more competitive in the scenario as petroleum resources deplete and costs rise. As with oil, increasing natural gas demands and regional variations in resources leads to increasing dependence on natural gas imports in many regions. As the scenario progresses, many regions become increasingly dependent on exports, especially from the Middle East and FSU. OECD regions' dependency on imports rises to nearly 50 percent by 2025 and grows still further after that. These patterns of supply dependence have similar implications as discussed above for oil, although the situation is less urgent. Non-conventional sources of natural gas could ease the dependency. For example, hydrates (icy substances composed of methane and water in deep-sea sediments) are a potentially enormous source of energy, but are economically and environmentally uncertain (WEC 1995). Coal reserves remain abundant despite the increasing demands in the scenario; environmental concerns are the limiting factor on coal use.

Nuclear energy

The long-term role of nuclear power in addressing the climate problem and a transition to sustainable energy use is a key uncertainty for the future. The nuclear power industry enjoyed an astonishing boom from its inception in the mid-1950s to the late 1980s. In 1995, nuclear reactors accounted for 17 percent of world electricity production. But in recent years the industry stagnated, with new construction starts barely keeping pace with plant retirements. Both the rising costs of nuclear energy and public concerns about nuclear safety have plagued the industry and brought it to a virtual standstill.

That stasis ends in *Market Forces* as nuclear power construction resurges. In the context of growing world electricity demand and increasing concern about global climate change, pressure is mounting already to turn to the nuclear option, a relatively abundant and virtually CO_2-free resource (OECD 1997). In the scenario, global electricity demand grows by a factor of 4 between 1995 and 2050, and more than a factor of 10 in developing regions. Installed generating capacity more than doubles by 2050. Much of the expansion is in developing regions, where nuclear power supplies 4 percent of power today but 14 percent by 2050.

Despite heady forecasts in the early years of the industry that nuclear power would be "too cheap to meter," it has proven in practice to be too expensive to build in most places. In the United States, construction costs soared to meet new safety regulations (Flavin 1984). The scenario assumes the introduction of a new generation of less costly, more modular and more reliable technologies, and improvements in conventional designs. At the same time, the rising costs of fossil fuel alternatives – and the continuation of various subsidies for nuclear power – would set conditions that are favorable for market reentry.

To address the sustainability of such a global resurgence of nuclear power, we consider three types of problems: safety, radioactive waste disposal, and international security. The early debate was sharp and inconclusive about the accident risks at a nuclear facility (Lovins and Price 1975, US Nuclear Regulatory Commission 1975, von Hippel 1977). Whatever the theoretical risks, the public reaction to the incidents at Three Mile Island and Chernobyl had a chilling effect on the industry. The expansion of nuclear power assumes that safety concerns are muted by demonstrated reliability and safety of advanced designs. However, the sustainability of nuclear power will continue to be compromised by the risk that a major accident in the future, as in the past, could prompt, calls for moratoria on nuclear power.

Nuclear generation produces prodigious quantities of highly radioactive waste from spent fuel, about 10,000 t per year globally. The waste contains highly toxic plutonium 239, with a half-life of some 20,000 years. The global accumulation in the scenario would grow from less than 100,000 t now to over a million tonnes by the year 2050. The decommissioning and dismantling of power plants at the end of their useful life would add significantly to radioactive waste inventories. In anticipation of this problem and feared shortages of uranium resources, the original concept was a closed nuclear fuel-cycle in which plutonium and uranium would be recovered, separated, and reused. But technical, cost and environmental difficulties have made the open-cycle once-through technology the norm, leaving much of the radioactive waste to accumulate in storage pools at nuclear plant sites. But as the United States has found, finding an acceptable site and design for a permanent repository is exceedingly difficult, due to technical uncertainties and immense local resistance. The need to monitor and contain radioactive and long-lived waste for thousands of years is a legacy to future generations that is difficult to reconcile with sustainability tenets.

Finally, the most vexing problem is the link between nuclear generation and security. The development of civilian nuclear power has contributed to the parallel development of nuclear weapons in several countries. The appearance of Russian plutonium on the black market of Europe and the Middle East raises the specter that terrorists or aggressive states will have increasing access to the raw material for nuclear weapons. The expansion of global commerce in fissionable materials, as envisioned in the scenario, would increase the dangers.

The high radioactivity of spent fuel from conventional open-cycle reactors mitigates the risks, since a sophisticated reprocessing facility would be needed for extracting plutonium. The closed-cycle technology, in which spent fuel is reprocessed and plutonium extracted, exacerbates the difficulties since reprocessed plutonium is sold on global markets. As the number of countries engaged in plutonium commerce expands, and the enforcement of global monitoring and safeguards becomes more difficult, the risk of nuclear terrorism rises. The increased security threats could compromise the institutional stability of the scenario.

Toxic waste

Toxic wastes pose particular threats in the environment because of their toxicity, persistence, and tendency to concentrate as they move through the food chain and accumulate in living organisms. Tracking the tens of thousands of potentially toxic chemical compounds exuded by modern industrial systems is intractable. However, the World Bank has devised an "Industrial Pollution Projection System" based on US data, which provides lower-bound toxics "emission factors" for many industries, expressed as emissions per unit of value added in each industrial subsector (Hettige *et al.* 1994). Applying these emission factors to industrial output for 1995 yields aggregate emissions of approximately 28 Mt in 1995. Figure 4.8 shows how this breaks down by industrial subsector and region.

The chemicals industry accounts for the largest share (44 percent), with metals industries also significant. The regional picture is dominated by the OECD regions, which account for 43 percent of all toxic releases. These estimates are for illustration only, due to the uncertainties in the US data itself and in applying them to other regions. In general, the effect will

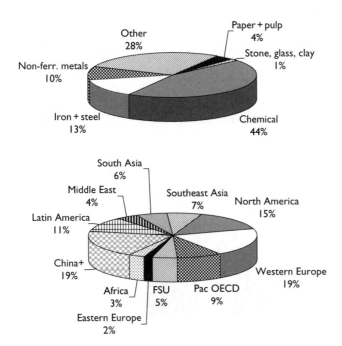

Figure 4.8 Current toxic waste emissions.

be to underestimate actual emissions in less-developed regions, because technological efficiencies and environmental controls are likely to be inferior (Jackson and MacGillivray 1995).

The current and future patterns in *Market Forces* are shown in Annex P-11. Based on the policy-as-usual premise of the scenario, emission factors per unit of value added are held constant over the scenario period, on the assumption that current levels of regulatory control persist in developed regions and that developing regions gradually converge toward these practices. Toxic emissions by 2050 are about three times higher than current levels. There is a roughly fourfold increase in the developing regions as a whole and a 70 percent increase in the OECD regions. These increases are somewhat less than the growth in the global economy, due to structural shifts in outputs and the playing out of current policies.

In the *Market Forces* scenario, international and national initiatives to limit waste production are unsuccessful. The "clean production" revolution required to counteract the growth in the scale of production and consumption requires a different global development trajectory. Our nominal sustainability goals – 50 percent reduction of toxic waste generation in OECD regions by 2025 and 90 percent by 2050, with convergence toward lower emission intensities in developing regions – are violated.

Water resources

In the *Market Forces* scenario, the pressure on freshwater resources grows more severe in all regions. The situation in Middle East countries – which is already grave – deteriorates, while problems deepen in Eastern Europe, China, Africa, South Asia and in various river basins elsewhere. Despite the much greater water-use efficiencies assumed in the scenario, population and economic growth drive up overall requirements.

The evaluation of water sustainability for the scenario has been conducted on a national basis, as regional analysis is far too aggregate for water assessment and can mask problems at the basin level. For each country, the population in water stress is assumed to rise from 0 to 100 percent as the use-to-resource ratio increases from 0.1 to 0.4, with high stress defined as a use-to-resource ratio greater than 0.4 (Raskin *et al.* 1997). Regional water stress indicators are summarized in Annex P-3, while regional and sectoral breakdowns of water withdrawals are summarized in Annexes P-1 and P-2, respectively.

In 1995, an estimated 1.8 billion people lived in countries experiencing some water stress (use-to-resource ratio greater than 0.1) and of these, over one-half billion were in

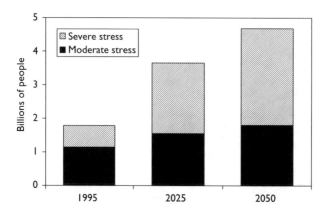

Figure 4.9 Population in water stress.

severe-stress areas. In the scenario, the population experiencing some water stress rises to about 4 billion by the year 2025, with 2.1 billion under high stress. In 2050, nearly 5 billion people contend with water stress, with 3 billion of them, over half of the total, in areas of high stress (Figure 4.9). The Middle East, already short of water today, faces a deepening water crisis, while huge populations in the Asian regions move toward high water stress levels, and Africa contends with increasingly intense water "hot spots."

Competition for scarce freshwater resources would hamper attempts to increase access to safe drinking water and sanitation. Beyond this immediate distress, water scarcity can also pose increasing security risks in shared water basins, such as in the Middle East (Wallensteen and Swain 1997). A highly uncertain, but potentially important, consideration is that the climate change impacts of the scenario may exacerbate the water situation in vulnerable areas (Morita *et al*. 1995, Alcamo *et al*. 1997). Far from meeting our sustainability criteria, water scarcity may be a critical limiting factor on the robust population and income growth of the *Market Forces* scenario.

Land resources and biodiversity

The requirements for cropland, grazing land, and human settlements place considerable pressure on existing land resources. Land degradation adds to this pressure since it leads to reduced productivity and even the abandonment of previously fertile areas. Lost farm outputs must be replaced by bringing additional land into production. As human populations and economic activity increase in the scenario, so does competition over land to support agriculture, forestry, development, recreation, and nature.

To track land change in the environment, we separate total land area into ten land-use types (Figure 4.10): built environment; cropland; grazing land; four different categories of

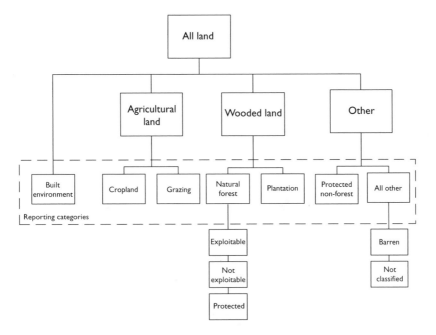

Figure 4.10 Land-use categories.

Table 4.1 Land areas and conversion in the Market Forces scenario

Land use	Area in 1995	Share of area in 1995 (%)	Increase (decrease) in conversions to (from)				Area in 2050	Share of area in 2050 (%)	Average rate of change (Mha/year)
			Built environment	Cropland	Grazing land	Forest			
OECD	**3,176**	**100**					**3,176**	**100**	—
Built environment	76	2	27	—	—	—	103	3	0.5
Cropland	406	13	(5)	34	—	—	434	14	0.5
Grazing	776	24	(6)	(21)	6	—	755	24	-0.4
Natural forest	1,043	33	(7)	(16)	(1)	(60)	959	30	-1.5
Plantation	41	1	0	0	0	71	112	4	1.3
Protected (non-forest)	283	9	0	0	0	0	283	9	0.0
All other	551	17	(9)	4	(6)	(11)	530	17	-0.4
Transitional	**2,284**	**100**					**2,284**	**100**	—
Built environment	16	1	3	—	—	—	19	1	0.1
Cropland	268	12	(1)	40	—	—	307	13	0.7
Grazing	370	16	(2)	(35)	57	—	390	17	0.4
Natural forest	813	36	(0)	(12)	(12)	(9)	781	34	-0.6
Plantation	23	1	0	0	0	10	33	1	0.2
Protected (non-forest)	8	0	0	0	0	0	8	0	0.0
All Other	786	34	(0)	7	(45)	(2)	745	33	-0.7
Developing	**7,561**	**100**					**7,561**	**100**	—
Built environment	175	2	285	—	—	—	460	6	5.2
Cropland	829	11	(82)	179	—	—	925	12	1.8
Grazing	2,261	30	(82)	(179)	620	—	2,620	35	6.5
Natural forest	2,213	29	(61)	(156)	(472)	(51)	1,474	19	-13.4
Plantation	39	1	0	0	0	66	105	1	1.2
Protected (non-forest)	278	4	0	0	0	0	278	4	0.0
All other	1,766	23	(60)	157	(149)	(15)	1,698	22	-1.2
World	**13,021**	**100**					**13,021**	**100**	—
Built environment	267	2	316	—	—	—	583	4	5.7
Cropland	1,503	12	(88)	252	—	—	1,667	13	3.0
Grazing	3,406	26	(90)	(235)	683	—	3,764	29	6.5
Natural forest	4,069	31	(68)	(184)	(484)	(119)	3,213	25	-15.6
Plantation	103	1	0	0	0	147	250	2	2.7
Protected (non-forest)	570	4	0	0	0	0	570	4	0.0
All other	3,102	24	(69)	167	(199)	(27)	2,973	23	-2.3

Note: All areas in millions of hectares (Mha).

forest and wooded land (commercially exploitable, not commercially exploitable, planta-tions and protected forest); non-forested protected land; barren land; and a residual "other" land category, which might include, for example, grasslands not used for grazing. The "reporting categories" shown in the figure are used to summarize results.

The changing pattern of land use in the scenario is summarized in Annex P-8. More detailed results for the conversion between different land uses is gathered in Table 4.1. The dominant change is from forests to settlements and grazing land. The loss of forest is particularly pronounced in developing regions. The dynamics of land change are complex, depending on settlement patterns, agricultural practices, economic growth, and natural resource industries. The key factors that drive change in the *Market Forces* scenario are discussed below.

Protected land

In many countries, land has been set aside for environmental protection, scientific research, education, and maintenance of traditional cultures. About 7 percent of all global land is protected, with regional figures reported in Table 4.2 (FAO 1998b, WCMC 1998b). Protected land in OECD regions is higher than average, at nearly 10 percent, with Southeast Asia, where 12 percent of total area is protected, an exception to the lower rates in non-OECD regions. The scenario assumes no net change in protected areas.

Settlement area

The human footprint on the landscape expands with development. In both urban and rural settings, land is cleared and altered for businesses, residences, roads, parking lots, landfills, burial grounds, and a host of other uses. This *built environment* expands onto agricultural land, forests, and other land types as populations grow and economies modernize. The built environment is still small, comprising only 2 percent of the world's land area. But it is growing and the impacts are costly to reverse.

Table 4.2 Protected land

Region	Total protected (Mha)	% Forest	Total protected as % of regional area
North America	187	15	10
Western Europe	45	22	9
Pacific OECD	102	13	12
FSU	70	94	3
Eastern Europe	7	44	8
Africa	149	23	5
China+	68	21	6
Latin America	129	38	6
Middle East	19	0	3
South Asia	20	49	5
Southeast Asia	52	100	12
World	850	33	7

An important and visible component of the built environment is urban land. In most regions where figures are available, urban area is expanding more rapidly than urban population, as land area per person increases (Douglas 1994). As the urbanization surge proceeds in the scenario, expansion of the built environment will be an important factor determining future changes in land use. Much settlement expansion is at the expense of potential agricultural land (FAO 2000d). The loss of arable land to expanding cities contributes to the problem of sustainably meeting growing agricultural requirements in the scenario.

In *Market Forces*, population and economic growth, cultural preference and land constraints condition changes in settlements. Total built areas can be decomposed into the product of two factors, area per capita and population. Estimates of current settled area per person are shown for regions in Table 4.3. Regions with relatively low incomes and high population density, such as China, South Asia and Southeast Asia, have relatively low values. Regions with higher income and high population density, such as Europe, have built environment per capita higher than in the Asian regions, but smaller than in North America. Finally, in Africa and Latin America, where land is plentiful but incomes are relatively small, built environment per capita is comparable to that of Western Europe.

Future *Market Forces* built environment trends in OECD regions are based on the recent experience in the US, where growth in developed land area exceeded the rate of population growth by about 0.6 percent per year. In the scenario, built environment per capita increases in all OECD regions at half this rate, gradually approaching 0.15 and 0.07 ha per capita in North America and the other OECD regions, respectively.

The land-abundant developing regions of Africa and Latin America converge gradually towards OECD average as incomes grow. The relatively land-scarce regions of China+, South Asia, Southeast Asia and Eastern Europe converge towards the more compact value for Western Europe and OECD Pacific. Although land is abundant in the Middle East and FSU, their built environment per capita is assumed to converge toward the compact European settlement pattern, rather than to levels more characteristic of North America. While land

Table 4.3 Current accounts built environment per capita

Region	Built environment per capita (ha/cap)	Source
North America	0.13	c
Western Europe	0.06	b
Pacific OECD	0.06	d
FSU	0.04	d
Eastern Europe	0.04	b
Africa	0.07	a
China+	0.03	a
Latin America	0.05	a
Middle East	0.06	a
South Asia	0.03	a
Southeast Asia	0.03	a

Sources: (a) based on Fischer (1993); (b) based on WRI (1994); (c) "built-up land" (USDA 1994) less state parks (USBC 1992, 1995); (d) based on other regional values.

is plentiful in the Middle East, water is not; the costs of expanding water delivery systems are likely to keep settlement densities relatively compact.

These patterns would place significant pressure on agricultural lands and valued ecosystems. The increase in the built environment globally in the *Market Forces* scenario is some 300 Mha, accounting for nearly one-half the total loss of forestland (Annex P-7). In developing regions, built-up areas more than double, eating into prime farmland, forests, and wetlands. The densely populated South Asia region loses significant swaths of both agriculture land and forest to land development. In convergent economic development, human settlement and land-use patterns approach those of highly developed countries. This basic premise of a *Market Forces* world implies the loss of ecosystems and weakening of agricultural productivity.

Agricultural land

Competing forces drive the need for agricultural land in the scenario. On the one hand, more land is needed in response to higher food demands. On the other hand, less is needed as crop and livestock productivity increase. The net effect is a gradual expansion of farms and grazing land, from 38 percent of total land area in 1995 to 42 percent by 2050, or about 10 Mha per year. The pressure to increase farmland is particularly strong in developing countries, as they seek to expand food production while maintaining historic levels of self-sufficiency. In the aggregate, developing-region agricultural area increases from 41 to 47 percent of total land area.

New cropland comes from the conversion of potentially cultivable land (PCL) that lies under forest, pasture, and rangeland. Total PCL per capita (including active and potential agriculture lands) provides a rough indicator of arable land abundance in a country. Regional values range from over 2 ha per capita in fertile and sparsely populated Latin America, to less than 0.3 in densely populated Asia (Annex F-8). In the scenario, cropland in the developing regions expands into pastureland and noncommercially exploitable forests in proportion to the amount of potentially cultivable land currently under those uses. In land-scarce regions, cropland expands also onto marginal land, which will require considerable inputs and careful management if it is to be sustainable.

Land degradation

Unsustainable land-use practices can lead to various forms of land degradation including wind and soil erosion, waterlogging, salinization, and nutrient depletion. Global land degradation of various degrees of severity has been estimated at 2 Gha in the last 50 years, or nearly a quarter of agricultural land, forest, woodland, and pastureland (Oldeman *et al*. 1991). The severity of land degradation ranges from diminished yields to severe deforestation. The quantity of land so severely degraded that it is removed from agriculture production has been estimated at 5–10 Mha per year (Kendall and Pimentel 1994), though the basis for these estimates is weak and they may be high. In the *Market Forces* scenario, about 3 Mha per year are withdrawn from agricultural production globally due to severe land degradation. This is cumulatively significant, amounting to 10 percent of current agriculture land by 2050. Since the net increase of global agricultural land is 10 percent (Annexes P-7 and P-8), the gross expansion is about twice that to compensate for agricultural land lost to degradation.

Forests and forestry

From 1980 to 1995, the average rate of forest loss worldwide was about 12 Mha each year (FAO 1997b). In the developing world there was a net loss of some 200 Mha over this period, while net reforestation and afforestation in the industrialized countries added about 20 Mha. Forest area loss erodes many valued services: forests are believed to contain two-thirds of the world's species; they are home to forest-dwelling people; they provide a diverse supply of products; they are a source of income for the poor; and they help mitigate global climate change.

Under *Market Forces* conditions, the historic dynamics driving deforestation persist. Expanding human settlements and agricultural land, and increased forestry pressure, combine to alter regional land cover. The cumulative loss of area under natural forest is over 800 Mha during the 1995 to 2050 period. This implies a nearly 20 percent loss of existing forest stands. Forest area in the scenario is reported in Annexes P-7 and P-8.

An important source of pressure on forested land is the rising demand for wood and other forest products (Annex P-6). Demand for paper products has grown particularly rapidly, especially in developing countries, both for publishing and packaging (Matthews and Hammond 1999). Recycled paper and other non-forest sources provide a significant proportion of fiber for paper production. But the bulk of paper production continues to rely on wood products.

The rising demand for wood has driven the expansion of forest plantations. While plantations constitute only a small proportion of total forest area in most regions, they are growing rapidly and are already quite significant in China+ (Annex P-6). Recent estimates indicate that the area under plantations is increasing at 4 Mha per year in tropical and subtropical countries, or roughly one-third the rate of tropical deforestation (FAO 2000c). Indeed, much of the expansion of plantations was from converted natural forest (FRA 1996). Forest plantations yield more fiber per hectare than natural forests, but they are far less biologically diverse. The lost ecosystem services are rarely reflected in the costs and benefits of forestry.

In the scenario, global requirements for wood and other sources of fiber more than double between 1995 and 2050. Around 60 percent of the 3.7 Gm^3 required in 2050 is supplied from forests, less than the 80 percent supplied in 1995 but more in absolute tonnage (Annex P-6). The remaining requirements are supplied from recycled and non-wood fiber sources, as well as from trees outside of forests. An increasing amount of production is met from plantations, which grow by nearly 1 Mha per year between 1995 and 2050 (Annexes P-6 and P-8). Consistent with past trends, most new plantation area is converted from previously forested land.

Natural forest areas decrease in the scenario due to conversions to plantations and other land uses. Comparing wood production to potential production on natural and plantation forest, the level of exploitation of the world's forest lands increases from less than 50 percent in 1995 to nearly 60 percent by 2050, with higher levels of exploitation in some regions (Annex P-6). These trends suggest that the historic pattern of unsustainable forest use would continue in a *Market Forces* global development scenario.

The scenario presents a picture of increased pressure on resources and the environment. If the environmental criteria for sustainability are to be met, policies will be needed to induce alternative patterns of resource use and to stimulate the development and deployment of better technologies. The kinds of adjustments necessary to meet environmental goals will be taken up in the *Policy Reform* discussion (Chapter 5).

Persistent poverty

The *Market Forces* scenario also is problematic with respect to social goals. Poverty reduction in the scenario relies on aggregate economic growth rather than targeted social policy. But is growth enough? Rapid increase in average income, indeed, will tend to reduce the incidence of poverty. But aggregate economic growth is only one of several interacting factors that determine the number of people in poverty. Population growth in poor regions and the drift toward unequal distributions of income are important countervailing effects.

National income distributions are typically skewed, with most of the population concentrated at lower incomes. The distribution of income is conveniently represented by the *Lorenz curve*, which plots the fraction of total income held by a given fraction of the population, beginning with the lowest-income populations (Figure 4.11). In a society where everyone has the same income, the Lorenz curve would be a straight line, while in real countries the curves take a concave form similar to that shown in the figure.

The *Gini coefficient* provides a useful summary indicator of the degree of inequality in a given society. It measures the deviation of the actual income distribution from a condition of perfect equality. With reference to Figure 4.11, the Gini coefficient is defined as the ratio of the areas $A/(A+B)$. The coefficient ranges from values of zero (absolute equality) to one (extreme inequality). Also shown in the figure are the *quintiles*, the fraction of total income held by the lowest-earning 20, 40, 60 and 80 percent of the population. In addition to the Gini coefficient, another useful measure of economic inequality, and one we shall make use of here, is the "low/high" ratio – the average income of the lowest-earning 20 percent of the population to the highest-earning 20 percent. The low/high ratio is defined as (Quintile 1)/ (1 − Quintile 4).

Income inequality varies widely across nations, the juridical unit at which international data is reported (Deininger and Squire 1996, Tabatabai 1996, USBC 1997, UNU/WIDER

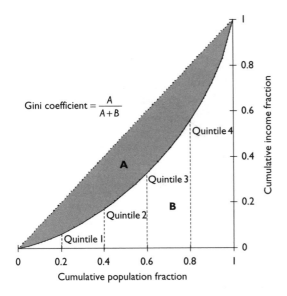

Figure 4.11 The Lorenz curve and the Gini coefficient.

1999, and World Bank 2000). Gini coefficients and low/high ratios are presented in Annex S-1 by region, where regional values have been computed by averaging national coefficients weighted by population. The regional Gini coefficients range from about 0.30 to over 0.50.

Income distributions in the *Market Forces* scenario tend to converge, as development patterns of poorer regions gradually approach those of industrial countries. In the context of global economic competition and weak social policies to address poverty and inequality, income distributions tend to become less egalitarian almost everywhere. In the scenario, the laissez-faire philosophy that prevails in the United States gradually spreads to other countries, as they seek to lower labor costs and weaken social safety nets in order to compete in global markets.

For 30 years, inequality has been increasing in the United States, as shown in Figure 4.12. In recent decades it has increased also in a number of Western European countries (Ruiz-Huerta *et al.* 1999). In the scenario, income inequality continues to increase in the United States, but at roughly half the historic rate as the shifts in wage distribution gradually stabilize. The Gini coefficient in the US reaches 0.50 in 2050, with other countries gradually converging toward this pattern (Annex S-1). The general trend is one of increasing Gini coefficients, with only Latin America showing a slight decline from the high current levels of inequality.

For each country and region, increasing populations, growing economies, and changing equity combine to alter the shape of income distributions. This is illustrated for the United States and China in Figure 4.13.

Population, income and distribution determine future poverty in the scenario. Poverty is often defined by a *poverty line*. When incomes (or expenditures in some surveys) are below

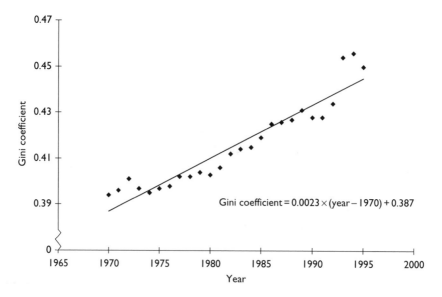

Figure 4.12 Historical US Gini coefficients.
Source: USBC (1997).

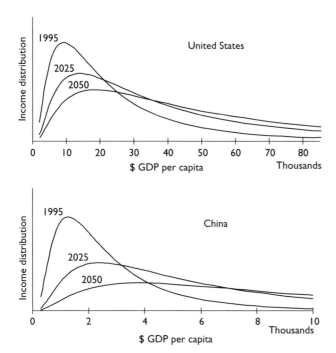

Figure 4.13 Changing income distributions.

this line a person (or household) is barely able to satisfy basic needs requirements. Since these lines are often set politically, we shall use a different measure – chronic undernutrition ("hunger") – as a proxy indicator for poverty, since its definition is more objective and less subject to cross-national variations in definition.

Correspondingly, we can define a *hunger line*, a threshold income below which an individual or family is unable to obtain the calories required to sustain a normal level of activity. Note that the hunger line includes more than the cost of food, since clothing, shelter, and fuel must also be obtained even by those just able to meet their basic nutritional needs. The hunger line is an approximation since some people with incomes above the line will suffer from hunger and some people below the line will be adequately fed. Also, since GDP per capita is not identical to income, hunger lines do not literally represent threshold income levels. Nevertheless, consistent use of the hunger line method in the scenario analysis accurately tracks changes of hunger levels from base year levels. The relationship between the number of hungry people in a region and the hunger line is shown in Figure 4.14.

The hunger line, the number of people hungry and the income distribution $p(i)$ are related by

$$\# \text{ hungry} = \int_0^{\text{hunger line}} \mathrm{d}i \, p(i).$$

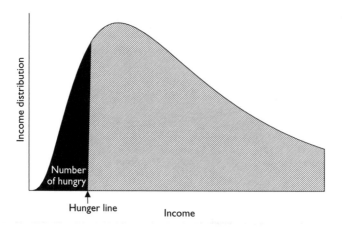

Figure 4.14 Relationship between number of hungry and the hunger line.

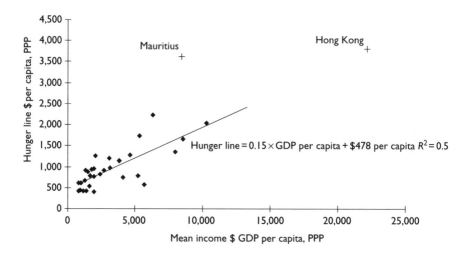

Figure 4.15 Hunger lines vs mean income.

Source: Current hunger levels from FAO (1997a); Incomes from WRI (1996b). Hong Kong and Mauritius excluded from regression. See Kemp-Benedict *et al.* (2002) for details.

Current values of national hunger lines are computed from this equation using income distributions and survey data on the number of hungry people. The hunger line cut-off tends to increase as countries develop (Figure 4.15). The increase is analogous to the well-known observation that poverty lines tend to rise as average incomes do (World Bank 1990, Ravallion *et al.* 1991). The increase in hunger lines may be due to decreased access to informal sources of food as countries modernize, a form of income that is not well cap-tured in household surveys. Furthermore, as countries grow wealthier, more must be spent on non-food items even by those just able to acquire the nutrients they need for a normally active life, in part because it takes a larger income to barely survive as traditional sources of material support are eroded in the process of modernization. The minimum income needed to avoid hunger is higher in Manhattan than in New Delhi.

Future national hunger levels in the scenario are computed as a function of changing population, income distribution, and the hunger line value in each region. The regional results are reported in Annex S-2. The number of undernourished people declines slightly over the course of the scenario, from over 800 million currently, to close to 700 million in 2050. The decrease in hunger to 200 million by 2050 called for by the sustainability targets cannot be expected if the trends in distribution, population, and economies captured in the *Market Forces* scenario persist into the future.

Regional variation of the incidence of hunger in the scenario is significant (Annex S-2). Africa shows the sharpest decrease in the percentage of population that is hungry, from nearly 30 percent in 1995 to 13 percent in 2050, but by far the largest increase in absolute numbers. In China+, South Asia and Southeast Asia, hunger decreases, while in other developing regions there is some increase. In the transitional regions of the FSU and Eastern Europe, the figures rise as relatively equal income distributions tend toward industrial country patterns.

While the quantitative analysis has focused on hunger, similar factors govern the behavior of other social indicators in a *Market Forces* world – unsafe water, illiteracy, and life expectancy. Economic growth will tend to improve the conditions of the poor, while growing populations and greater inequity are counteracting. In general, the patterns for these indicators will be similar to those for nutrition. As a percentage of population, the ranks of those who are illiterate or lack safe water will decline. But the goals of significant reductions in absolute number will not be achieved.

Conclusions

Measured against our sustainability criteria, the *Market Forces* scenario would be a risky bequest to future generations. The increasing pressure on environmental systems is environmentally unsustainable. While there is great scientific uncertainty on how ecological systems would respond to increasing pressure, and where thresholds of sudden change might lie, the scenario would flirt with major ecosystem state changes and unwelcome surprises. Environmental feedbacks and impacts on human systems could undermine a fundamental premise of the scenario – perpetual economic growth on a global scale.

The scenario also fails to address the social goals for sustainability. Absolute poverty persists, as 700 million people remain hungry at mid-century. Beyond failing the ethical imperative of sharply reducing human deprivation, social conditions would be maintained that link human desperation and environmental stress. The persistence of economic polarization could compromise social cohesion and feed alienation and anger. Social and political tensions would be magnified by deepening resource and environmental pressures – conflict over water, regional concentration of petroleum supplies, scarcity of land, climate change impacts on land and water resources, biodiversity loss. The spread of the liberal democratic institutions that are assumed in the scenario would become more difficult and their maintenance more fragile.

In addition to greater disparities within many countries and regions, the absolute disparity between rich and poor countries increases, though income grows fastest in poorer regions. The desire to migrate to rich areas would grow stronger, as would the resistance to such migration. Interregional inequity also could aggravate geopolitical tensions. The fundamentalist backlash to the process of global cultural homogenization, spurred in the scenario by advancing markets and the information revolution, would be reinforced. The seedbed for violent extremism would be fertilized.

These social, political frictions, in the context of progressive environmental degradation, could nullify the continuity assumptions of the scenario. If allowed to fester, a xenophobic and isolationist *Fortress World* mentality could flourish in privileged areas – and the United Nations would have a weakened capacity to mediate such tensions in the market-driven world of the scenario. The danger would grow that the path of history would branch toward some form of *Barbarization* scenario.

The risks are real, but they can be moderated. The tensions inherent in a *Market Forces* scenario can be relaxed if development becomes more sustainable. Ultimately, the fundamentally new social arrangements and values of a *Great Transition* may emerge. In the near term, the transition to sustainability can be substantially advanced through the incremental adjustments within a *Conventional Worlds* paradigm. We turn now to a consideration of the scope for such *Policy Reform* scenarios.

Chapter 5

Bending the curve

Can a successful transition to sustainability be achieved within the confines of *Conventional Worlds* development assumptions? The policy-complacent *Market Forces* scenario has presented a picture of widening international prosperity in the next century. But it is also a picture with disquieting elements – increasing environmental risk, resource scarcity, tenacious poverty, and social tension. These threats could well negate the scenario's underlying assumption of steady and globally convergent economic growth, providing a seedbed for the polarization and conflict that feeds the forces of *Barbarization*.

In this sense, *Market Forces* scenarios may be unrealistic as well as unsustainable. We ask next: What types of strategies could bend the curve of future development toward sustainability goals? Can the incremental adjustments envisioned in a *Policy Reform* scenario diminish the risk of destabilization, while providing an environmentally and socially resilient basis for global development?

Policy Reform scenarios

Policy Reform scenarios maintain the essential assumptions of the *Conventional Worlds* paradigm. The elements by now are familiar – the steady march of globalization, the gradual convergence of all regions toward the evolving model of development in industrial regions and progressive homogenization of global culture around the values of materialism and individualism. But there is a critical difference between the scenarios.

In contrast to *Market Forces*, *Policy Reform* scenarios assume the emergence of a popular consensus and strong political will for taking action to ensure a successful transition to a sustainable future. In this vision, solemn international agreements to comprehensive environmental protection, development and justice, such as those of the 1990s, are no longer largely rhetoric that does not significantly alter development priorities on the ground. Instead, they are translated into a vigorous array of global, regional, and national targets. Working together, governments craft and implement an effective set of initiatives, relying on such policy instruments as economic reform, market mechanisms, regulation, social programs, and technology development.

We constructed the *Market Forces* scenario as a forecast, beginning with existing global conditions and drivers of change, and examining how the future might unfold over time. A different method is required for formulating *Policy Reform* scenarios which are aimed at meeting certain conditions in the future; namely, the sustainability targets. Rather than a forecast into the future, the *Policy Reform* scenario is a normative scenario that is constructed as a backcast *from* the future (Robinson 1990). We begin with a vision of a desired future state and seek to identify plausible development pathways for getting there.

Whereas business-as-usual scenarios pose the question of the forecaster – "Where are we going?" – normative scenarios add the questions of the traveler – "Where do we want to go? How do we get there?" *Policy Reform* scenarios illuminate the requirements for simultaneously achieving social and environmental goals of sustainability. They clarify the strategies for bending the curve of development away from *Market Forces* and toward a sustainable future.

In a *Policy Reform* world, poverty reduction goals are realized through initiatives to increase the incomes of the poor. Both greater international equity among rich and poor countries and greater national equity within countries are critical to meeting these goals. In the scenario, increased international equity is aided by strategies that accelerate the convergence of developing and transitional regions toward OECD levels of development. National equity is improved through a policy focus on poverty reduction and more equitable distribution of income.

Meeting the environmental goals in the *Policy Reform* scenario requires dramatic adjustments in the use of resources, abatement of pollution, and protection of ecosystems. In the face of increasing population and rapid economic growth, the simultaneous satisfaction of environmental and social targets presents an exceedingly daunting technological challenge. Rising requirements for energy, water, and other resources must be counteracted by high end-use efficiency in transport, industry, agriculture, and households. Renewable energy, ecologically based agricultural practices, and integrated eco-efficient industrial systems must become the norm. To live within water and land constraints, some resource-poor developing countries must rely more heavily on food imports from OECD and transitional regions, which in turn would require a reversal of recent trends in industrialized countries of withdrawing land from agricultural production.

The illustrative *Policy Reform* scenario (Raskin *et al.* 1998) is guided by the provisional sustainability goals described in Chapter 3. In some ways, these social and environmental objectives are at cross-purposes. The economic growth required for poverty reduction and greater international and national equity tends to increase certain environmental pressures, those associated with greater affluence. At the same time, the environmental problems associated with poverty, such as the unsustainable use of marginal land by subsistence farmers, would abate. In addition, more rapid development in poorer countries would offer the economic wherewithal for greater attention to pollution abatement and long-range resource and environmental preservation.

The scale of the required transition is illustrated by comparing *Market Forces* and *Policy Reform* scenarios across selected social and environmental dimensions, as shown in Figure 5.1. A detailed quantitative compendium of the two scenarios is tabulated in the Annex. As we shall see, there are numerous interactions between the many aspects that comprise this world development trajectory. The comprehensive policy strategies in the scenario simultaneously affect whole clusters of issues.

We begin with a discussion of the conditions for meeting social goals, specifically the targets for the reduction of human deprivation. This logically precedes a discussion of environmental impacts since the demographic, economic, and equity assumptions for meeting the social goals set the scale of resource requirements and environmental impact in the scenario. Then, we go on to consider the improvements in technological, resource use, and production practices required by environmental goals.

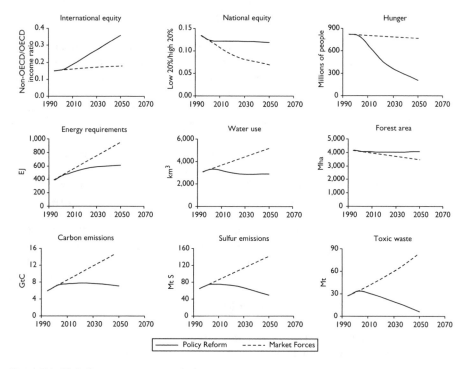

Figure 5.1 Global patterns compared.

Social goals

Poverty reduction strategies in *Policy Reform* scenarios reflect the market orientation of *Conventional Worlds*. The many dimensions of poverty are addressed through efforts to increase the incomes of the poor and their access to food and other basic needs. In principle, income can be augmented through welfare transfers, such as direct food aid, in addition to structural changes that provide the poor with greater access to income-generating activities. Indeed, targeted assistance programs for meeting basic needs would have high priority in *Policy Reform*. They address urgent needs and can contribute to self-sustaining processes that raise the incomes of the poor (UNDP 1997). However, a sustainable and resilient response to the challenge of poverty reduction must ultimately be reflected structurally in access to decent and secure livelihoods. In the scenario, this implies a strong policy focus on steadily increasing incomes of the poor everywhere over the coming decades through educational, financial, and social empowerment.

As discussed in the previous chapter, the level of hunger is used as a proxy indicator for poverty in general in the scenario illustrations. The methodology for determining hunger levels was described there: the number of the hungry in each country is determined by population, average income, and the distribution of income. The paths of these three factors in the course of the *Policy Reform* scenario are constrained by the sustainability goal of halving the hungry population between 1995 and 2025, and again between 2025 and 2050. They are also constrained by the requirement of consistency with the underlying *Conventional Worlds* story and the general goal of improved equity in *Policy Reform* scenarios.

Population growth in a *Policy Reform* development context would be somewhat less than in a *Market Forces* scenario. A number of factors – rising incomes of the poor, higher economic security, and more opportunities for women – would tend to reduce the desired size of families. Also, greater access to social services would increase control over reproductive capacity and reduce unwanted pregnancies. As the cycle of poverty is broken, the desire for fewer children would gradually moderate population increase. This would quicken the "demographic transition" toward stable populations at low birth and death rates.

The correlation between increases in income, education, and female empowerment, on the one hand, and decreases in fertility rates, on the other, is clear, but the detailed relationships are complex. Also, considerable momentum for future population growth is built into existing demographic structures as large cohorts of the young eventually pass through their childbearing years. In the scenario, we assume that this momentum is slowed, but only gradually. Population in non-OECD countries in 2025 is taken to be 98 percent of the UN mid-range projection, and in 2050 to be 95 percent of the UN value (see Annex D-1). This is a conservative, but reasonable, assumption given the great uncertainty on the drivers of population growth (Bongaarts 1997). The lower population levels help modestly in meeting the scenario goals of reducing poverty and environmental stress.

The other two factors influencing hunger levels – average income and income distribution – are related to two different dimensions of economic equity. The first is equity *among* countries. In *Policy Reform*, such *international equity* increases through concerted global strategies. These would include stimulation of investment opportunities in developing countries, strengthening of the enabling institutional climate for modern markets, and higher levels of direct financial transfers through official development assistance. Mechanisms built into international agreements on climate stabilization and other environmental issues support greater international equity by providing incentives for high-income countries to partially meet their responsibilities through investments in "clean development" in poorer countries. Initiatives such as these lead to more rapid economic growth in developing and transitional regions, and faster convergence toward OECD levels of development.

The second dimension of economic equity is equity *within* countries. In the scenario, such *national equity* also increases. The trend toward greater inequity, a trend that persists in *Market Forces*, is reversed. In *Policy Reform*, "growth with equity" becomes the prevailing philosophy in national development strategies. In this environment, human development indicators become as important as GDP growth in gauging national economic performance, commanding the attention of the public and politicians alike (UNDP 2001).

The challenge is to lift the incomes of the very poor above the "hunger line," the minimum income for avoiding hunger described in Chapter 4, in sufficient numbers to satisfy the sustainability targets. In principle, this could occur through various combinations of changes in the size of the global economy, international equity, and national equity. In the quantitative illustration of *Policy Reform*, the targets are met through a balanced combination – strong economic growth, gradual motion toward international equity, and far greater national equity than in *Market Forces* projections. The *Policy Reform* scenario satisfies the hunger reduction target of 50 percent by 2025 and an additional 50 percent by 2050, as shown in Annex S-2. The detailed economic and equity assumptions are discussed below.

In scoping the plausible range of economic growth rates over the half-century time horizon of the scenario, an examination of historical data provides guidance. For example, growth of the successful US economy over the last 30 years was around 3 percent annually. "Asian tiger" economies grew much faster over the first part of this period, by as much as 9 percent annually in some cases. However, the era of extremely rapid growth in Asia seems to have drawn to a close in light of recent economic downturns and instability. Taking a longer perspective, Thurow (1996) claims that "... over the past century, no country's economic growth rate has averaged better than 3.6 per cent per year."

With this in mind, a reasonable upper bound for national economic growth from 1995–2025 is one-half the growth rate of the fastest-growing Asian tigers, or 4.5 percent per year on average and 3.6 percent per year over the full 55 years of the scenario. As a lower bound, we assume that the economies in all regions grow, with GDP per capita growing at least 0.5 percent per year in OECD regions and 1.0 percent per year in non-OECD regions.

The range of possibilities is further restricted by the assumption of global economic convergence that underlies all *Conventional Worlds* scenarios; that is, international equity should not decrease over the course of the scenarios. With the ratio of average income per capita in non-OECD regions to that in OECD regions taken as a measure of international equity, the ratio would increase in the scenario from its current value of about 0.15. Perfect interregional equity would correspond to a value of 1.0.

The various constraints define an "envelope of possibility" that circumscribes a plausible range set of *Conventional World* scenario assumptions (see Figure 5.2). Also shown in the figure are the assumed values for *Market Forces* and *Policy Reform* scenarios. The scale of the global economy in the *Policy Reform* scenario is similar to that of the *Market Forces* scenario. It is in fact slightly smaller, consistent with the smaller global population in the *Policy Reform* scenario. However, the level of international equity is significantly higher in the *Policy Reform* scenario, while remaining well within the bounds set by the plausibility constraints.

Income is allocated within the OECD and non-OECD regions consistent with the assumption of rapid convergence of national economies. This pattern can be seen in the regional growth rates shown in Annex E-3. In Africa and South Asia, which have the lowest average income in 1995, income growth rates show the greatest increase in *Policy Reform*

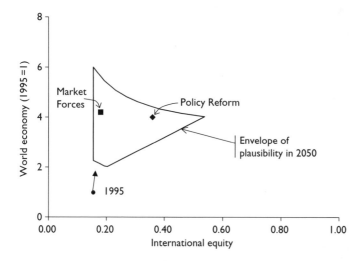

Figure 5.2 Envelope of plausibility.

relative to *Market Forces*. In contrast, China+, which experiences relatively high growth in the *Market Forces* scenario, and the Middle East, which has a relatively high average income in 1995, each show a smaller difference in income growth rates between the two scenarios.

The national equity variable is defined here as the ratio of the incomes of the lowest-earning 20 percent of a population to the highest-earning 20 percent within a country. Perfect equity within a given society would correspond to a value of 1.0. Actual values range widely across countries; for example, 0.14 in Denmark, 0.08 in the United States, and 0.04 in Brazil. Current average regional values are reported in Annex S-1, along with values for the illustrative scenarios. The gradual convergence of socio-economic patterns that is assumed in scenarios is reflected in the gradual convergence in income distributions over time, as well. As discussed in the previous chapter, *Market Forces* income distributions converge toward future US patterns, as the US income distribution widens but at slower than historical rates. In the *Policy Reform* scenario, changes in income distribution are constrained by the sustainability target of hunger reduction. They are shown in Annex S-1.

To clarify the dynamics of the scenarios further, it is convenient to introduce a single global variable describing the average pattern of national equity in a given scenario (see Annex S-1). The global average value of national equity (X) is computed as the population-weighted average of the national equity in each country (X_c). Thus, $X_t = \Sigma_c (P_{c,t} \cdot X_{c,t})/P_t$, where P_t is global population and the indices c and t signify country and year, respectively.

Our three critical variables – world economy, international equity, and national equity – define a three-dimensional space as shown in Figure 5.3. For a given year, a scenario appears as a point that moves through the space as the values of the variables change. In the course of time, each scenario defines a trajectory that departs from a common point of departure defined by current values of the world economy, international equity, and national equity. The "spheres of uncertainty" that surround each point in the figure illustrate a range of possibilities that are compatible with the basic scenario story lines.

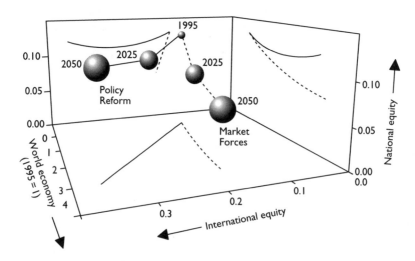

Figure 5.3 Scenario trajectories.

Notes: International equity is the ratio of average income in non-OECD regions to OECD regions. National equity is the global average ratio of the income of the poorest 20 percent to the richest 20 percent in each country, weighted by population.

In the *Market Forces* trajectory, the world economy quadruples by 2050, international equity improves gradually and national equity decreases. In the *Policy Reform* scenario, international equity increases substantially, from 0.15 in 1995 to 0.36 in 2050. Though the gap between rich and poor nations is far from closed, the level of international equity is twice that of the *Market Forces* scenario. National equity in the *Policy Reform* is nearly twice the *Market Forces* scenario value in 2050.

The decline in equity reverses and eventually returns to current values. Countries in regions that are particularly unequal today, such as Latin America, show sharp improvements in the convergence toward global norms (Annex S-1). By contrast, regions that are now relatively equal, such as the FSU and Eastern Europe, become less equal in the process of convergence, but much more equal than in *Market Forces*. In general, future equity assumptions are well within the bounds of recent historical experience. The scenario initiatives counteract the trend toward high inequity through policy reform, not extreme egalitarianism.

The link is strong between meeting hunger reduction targets and improved national equity relative to *Market Forces* trends. With the levels of inequality assumed in the *Market Forces* scenario, GDP_{PPP} in developing regions would have to increase at over 5 percent per year between 1995 and 2050, and at nearly 6 percent per year between 1995 and 2025 to meet the target. Such a high growth is implausible in light of the growth sustained historically over multiple decades. In addition, it would imply an immense expansion in the scale of the world economy – by a factor of nearly 15 by 2050 – with corresponding strains on environmental systems.

The challenges are to simultaneously foster world economic growth, development convergence, and greater national equity, while remaining within environmental sustainability goals. With the *Policy Reform* economic and distribution assumptions in hand for meeting social targets, we turn to the environmental aspects of the sustainability transition.

Climate

In a *Policy Reform* scenario, strategies for the energy sector are strongly influenced by climate change goals. To recap, the sustainability target for climate is guided by the principle that the rate of climate change should be sufficiently slow to allow ecosystems to adapt. We translated this goal into an upper limit on global temperature change of less than 2°C between 1990 and 2100. To achieve this, atmospheric concentrations of carbon dioxide need to stabilize at no more than 450 ppmv by the year 2100. This, in turn, implies cumulative anthropogenic carbon dioxide emission allowance of the order of 700 GtC between 1990 and 2100.

The issue of setting goals for carbon emissions has been the subject of considerable international negotiation and contention. In the first phase of the United Nations Framework Convention on Climate Change, the focus is on near-term targets for carbon reductions for industrialized countries, the Convention's so-called "Annex I countries," corresponding roughly to our three OECD regions and the two transitional regions of Eastern Europe and the FSU. These more industrialized countries have accounted for the bulk of carbon emissions historically. The Convention calls for them to take positive steps to reduce their carbon emissions as an initial phase toward a comprehensive global agreement. This is codified in the Kyoto Protocols to the Convention, which requires reductions that are differentiated across Annex I countries, but in the aggregate would reduce their carbon emissions by about 5 percent from 1990 levels by the 2008–2012 period (UNFCCC 1997).

Even if ratified by all Annex I countries, Kyoto is only a beginning in the process of fashioning an accord that bends the curve of global emissions from the high growth of *Market Forces* to the sharp reductions in *Policy Reform* (Annex P-4). To attain climate stabilization at safe levels, eventually all countries will need to adopt greenhouse gas emission constraints. How should the burdens be shared? What levels of allowable emissions are reasonable for each region and country?

An acceptable global agreement must allow for some emission increases in developing countries as their economies grow rapidly and converge toward OECD levels in the course of the twenty-first century. The eventual engagement of poor countries as active participants in a comprehensive climate stabilization agreement will need to recognize the right to develop. It also must reflect the historic responsibility of the developed world for experienced increases in the global concentration of carbon dioxide, which resides in the atmosphere for centuries, and other greenhouse gases. Political acceptability and simple fairness dictate that some formula for equitable emission rights must ultimately underpin and guide long-term climate policy.

These considerations are reflected in *Policy Reform*, a scenario defined by the emergence of a widespread governmental commitment to environmental sustainability and greater global equity. Several conditions determine regional emissions in the scenario. First, abatement targets that are strong but achievable are set for the OECD and transitional regions through the year 2025. This conforms to the Climate Convention principle of early and sustained action by industrialized regions. Second, in the long term, all countries approach a common per-capita emission allowance, following a long period of transition to emissions equity. Third, developing country emissions are increased in the transition period but are constrained to keep cumulative global emissions from all regions to the year 2100 at about 700 Gt.

The *Policy Reform* scenario illustrates how this strategy can be realized with specific assumptions for allowable emissions over time for countries and regions. The solution is not unique – alternative emission profiles over the coming centuries would satisfy the goals of climate stabilization and equity. Certainly, the definition of emissions equity will be the subject of debate in the years to come, as will the timing and mechanisms for achieving it. However, the scenario captures the essential contours of the class of all technologically and politically feasible solutions for rapid climate stabilization.

Relative to 1990 levels, OECD regions are assumed to reduce annual emissions by 35 percent in 2025. The pattern of emissions in transitional regions, where emissions dropped precipitously after 1990 as their economies shrank, is somewhat different. Their emissions rise until 2010 as their economies recover, then are reduced gradually to reach the 2025 target of 35 percent reduction from 1990 levels. These reductions in Annex I countries go well beyond those in the Kyoto Protocol, and are not on the radar screen of current official negotiations.

But the essence of *Policy Reform* is the kind of rapid shift in political will that would lead to vigorous action to mitigate the climate threat. The active participation by all major emitters is a prerequisite for success, especially the United States. Comprehensive programs that promote efficient technology, low-carbon energy sources and land-based carbon sequestration projects would need to be rapidly deployed. In practice, a portion of these reductions would be met through the "flexibility mechanisms" of Kyoto, which permit Annex I countries credit toward their reductions through investment in emissions mitigation projects outside their borders, particularly in developing countries. Environmentalists have pressed the need to design flexibility mechanisms that are compatible with the larger sustainable development agenda. For example, climate-driven forestry incentives should promote biodiversity protection, and energy strategies should stimulate sustainable livelihoods for the poor. These concerns would be more readily addressed in a *Policy Reform* context.

The assumption of eventual convergence of all countries toward common per-capita emissions (or "emission rights" in a tradable permit approach) addresses equity concerns. The immense disparities in per-capita emissions today (Annex P-4) are the result of many decades of divergent economic growth between rich and poor countries. They cannot be eliminated rapidly without politically unacceptable disruption. However, the goal of eventual equality is compatible with the assumptions in the *Policy Reform* world of eventual equity and convergence across regions. In the scenario, it is assumed that annual emissions converge to a common rate of 0.6 tC per capita in 2075, with equal per-capita emissions thereafter. This constrains emissions to about 3 GtC per year by 2100, given the mid-range population projections in the scenario, which is required to stabilize climate at the 450 ppmv concentration level. Finally, developing-country emissions increase substantially over the next decades, while remaining within the constraints imposed by the global cap on cumulative emissions of 700 GtC and the convergence target.

Figure 5.4 summarizes cumulative global emissions pathways of *Policy Reform* scenarios along with the *Market Forces* scenarios. The 1995 value includes estimates of total industrial emissions since 1850 from combustion of fossil fuels and cement production plus estimates of net emissions from land-use change. *Market Forces* shows an increase in cumulative emissions of 1,500 GtC. The *Policy Reform* challenge is to cut this in half, to less than 750 GtC. The third curve in the figure shows an emission profile leading to stabilization at 350 ppmv (IPCC 1995a), greater than the pre-industrial concentrations of 280 ppmv, and roughly equivalent to the current concentration. This "strong sustainability" alternative would allow for cumulative emissions of only about 220 GtC between 1995 and 2100, which is implausibly low under conventional growth assumptions, notwithstanding the action program of a *Policy Reform* world. It would take a *Great Transition* scenario to reduce the drivers of greenhouse gas emissions to that extent.

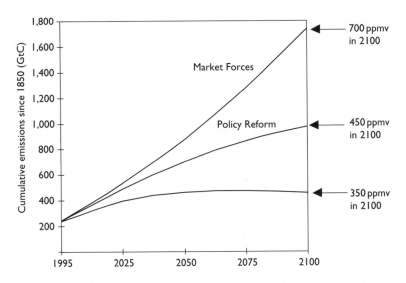

Figure 5.4 Cumulative global CO$_2$ emissions and atmospheric concentrations.

The emissions pattern for each of the three macro-regions, expressed as emissions per capita, is shown in Figure 5.5. In the developing regions, per-capita emissions start from a much lower level than in the OECD or transitional ones. They grow steadily until 2025, without ever exceeding the industrial or transitional rates, then begin to drop toward the 2075 convergence target. The global average of emissions per capita remains almost constant between 1990 and 2025, and decreases from 2025 to 2100.

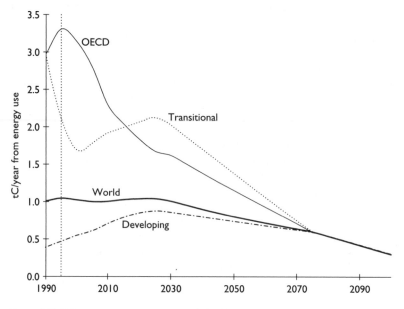

Figure 5.5 CO_2 emissions per capita in Policy Reform.

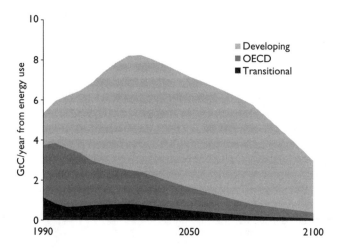

Figure 5.6 Annual CO_2 emissions in Policy Reform.

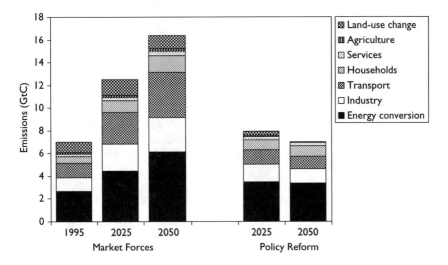

Figure 5.7 Annual global CO_2 emissions by source.

Combining emissions per capita with the population projections, we obtain total annual emissions in each macro-region shown in Figure 5.6. The total integrated emissions – the area under the curve in the figure – achieve the sustainability target of about 700 GtC. The share of emissions attributable to OECD countries falls from about 50 percent in 1995 to about 20 percent in 2050, as low carbon energy strategies moderate the emissions in the North, and population growth and accelerated economic development drive emissions higher in the South.

Applying the same approach, we assign emission goals at the regional (Annex P-4) and country level. OECD and transitional regions follow the general patterns outlined above, while emissions in the developing regions follow various paths, depending on current values and differing rates of economic growth.

The sectoral adjustments that lead to the climate stabilization targets are summarized in Figure 5.7, and discussed in the following sections. *Policy Reform* emissions are 60 percent lower than *Market Forces* in 2050 with decreases in all sectoral sources. The sectors that dominate emissions – energy production, transport, industry, and land change – are the sources of greatest policy effort. But initiatives to moderate greenhouse gas emissions must cut across the entire spectrum of human activity.

Energy

If there is to be a passage to an environmentally sustainable world, an energy transition will be at its foundation. The era of industrial development was based on the emergence of modern forms of energy – coal, oil, natural gas, and electricity – that were abundant, inexpensive and increasingly flexible. But energy has been a scourge as well as an engine for industrial society, implicated deeply in many of its environmental depredations – air and water pollution, acid rain, climate change, oil spills, toxic emissions, and many others.

The lesson of the *Market Forces* scenario is that gradual energy efficiency improvement and modest renewable energy growth is insufficient to meet sustainability goals (Chapter 4). Aggressive action will be needed on both the demand and supply sides of the energy equation. Two great pillars – efficiency and renewable energy – define the energy strategy of a *Policy Reform* world. On the demand side, the intensity of energy use (energy requirements per unit of activity) will need to be significantly moderated through the adoption of highly efficient energy use in vehicles, buildings, appliances, and industrial processes. On the supply side, energy must come increasingly from natural gas, the cleanest of the fossil fuels, during a transition period, while moving aggressively toward a world energy system based on solar power in its many forms. The sustainability agenda requires phasing down fossil fuels and nuclear power, and ramping up wind, biofuels, photovoltaics and other direct solar, and the full panoply of renewables sources.

Technologically and economically feasible energy pathways that can satisfy the stringent constraints imposed by climate stabilization and other environmental goals are available for each region. But the energy transition will require strong and sustained support for appropriate regulation, research and development and economic incentives, as envisioned in a *Policy Reform* scenario.

Energy intensity

Figure 5.8 illustrates how a sustainable energy transition might be realized. Global energy requirements increase from 392 EJ in 1995 to 600 EJ in 2050, a 53 percent increase from

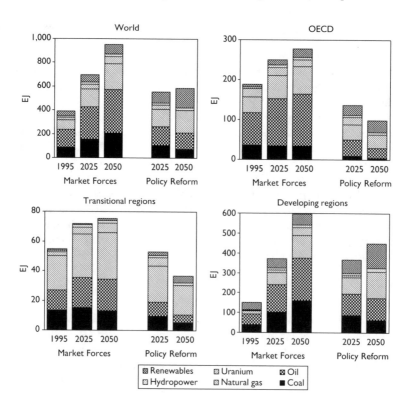

Figure 5.8 Primary energy requirements.

present consumption, but 37 percent lower than the 951 EJ in the *Market Forces* scenario (Annex En-1). In OECD regions, energy use declines by about 40 percent over this period, compared to an increase of nearly 50 percent in *Market Forces*. In the transitional regions, energy requirements remain relatively stable to 2025, increasing by 13 percent as improvements in energy efficiency are negated by a strong economic recovery. Developing-region energy requirements increase by a factor of over 3 between 1995 and 2050, which is less than the nearly fourfold increase in the *Market Forces* scenario, despite more rapid economic growth (Annex E-1).

Energy intensity changes for the *Policy Reform* and *Market Forces* scenarios are shown in Figure 5.9. Five major final energy-using sectors are analyzed for each region: industry, transport, households, services, and agriculture. The industrial sector is further disaggregated into five major energy-consuming subsectors. In the transport sector, freight and passenger transport are treated separately, with each broken down by mode (road, rail, air, and water).

In *Policy Reform*, the energy required per unit of economic activity decreases in OECD regions by 70 percent between 1995 and 2050. In developing regions, the decrease is closer to 60 percent. In transitional regions, where energy use is currently very inefficient, energy intensities decrease by over 70 percent between 1995 and 2050. The patterns are based on the rapid penetration of economically feasible measures in the OECD as the technological stock turns over. The other regions converge toward these values as their average incomes increase. This implies technological "leapfrogging" in the scenario, since developing regions approach best future practices for energy utilization, rather than replicating industrialized

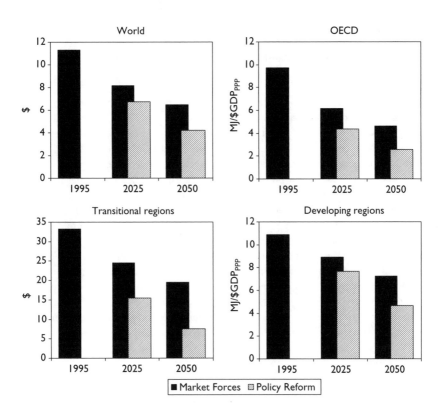

Figure 5.9 Energy intensities in the scenarios.

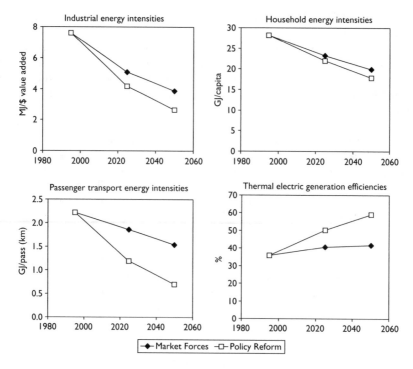

Figure 5.10 OECD energy intensity improvements by sector in the scenarios.

countries' practices in the course of development. The trends in energy intensity for OECD regions are shown at the sectoral level in Figure 5.10.

The *Policy Reform* scenario incorporates a wide variety of measures for achieving energy intensity improvements. In the residential and service sectors, these include improved space heating and cooling systems, more efficient appliances and lighting, better glazing and building insulation, greater use of heat pumps, and passive solar heating and cooling. Industrial sector improvements feature greater use of recycled material feedstocks, efficient motors, and combined heat and power technologies. In the transport sector, vehicle efficiency improves sharply. In North America, average mileage for automobiles increases from 19 miles per gallon (mpg) to 32 mpg in 2025 and 51 mpg by 2050. These improvements are well within the bounds of what is considered to be technically and economically feasible. For example, efficiencies in new automobiles could reach 75 mpg by that time (Energy Innovations 1997). In the electric power sector, a new generation of technologies such as combined-cycle thermal plants and integrated coal and biomass gasification provide substantial efficiency increases in the medium term. In the longer term, fuel-cell technologies and greater use of combined heat and power will be required to meet the sector efficiency goals of the scenario.

Fuel switching

The second prong of a sustainable energy strategy is switching from fossil fuels to a greater reliance on renewable energy sources, with natural gas as a near-term bridge fuel. The composition of energy supplies shifts fundamentally in the *Policy Reform* scenario driven by carbon emission targets (Figure 5.6). Similar patterns apply at the regional level (Annex En-2).

The combined share of the most carbon-intensive fuels, coal and oil, decreases from 60 percent in 1995 to 35 percent in 2050, while the share of natural gas increases from 20 percent in 1995 to 37 percent in 2050. This can be compared to the *Market Forces* scenario, in which the share of coal and oil remains almost constant. Severe pressure on fossil fuel resource availability is postponed but not entirely alleviated, as shown in Annex En-5. In addition, the scenario assumes the phase-out of nuclear generation, for reasons noted in Chapter 4. If the nuclear option were included, carbon emission and other environmental targets could be met with less stringent requirements for renewables and efficiency improvement.

The mix of renewable energy forms varies with region according to natural resource endowments – hydroelectric potential, insolation, availability of land, and the presence of suitable water and soil conditions for modern biomass plantations. Renewable options include biofuels as a feedstock for the industrial, transport and electricity generation sectors, solar energy for electric generation and for direct applications in households and industries, and wind and geothermal energy for electric power generation. Hydropower doubles, as it does in the *Market Forces* scenario, with more emphasis on smaller projects that avoid the environmental disruption and human displacement of many large hydro schemes.

Levers of change

The efficiency and fuel-switching initiatives in the scenario require neither heroic technological assumptions nor economic disruption. Indeed, the promotion of energy efficiency and renewables is not only compatible with economic development and job stimulation; in fact, it may further these goals (Energy Innovations 1997). The constraints to the energy transition are institutional and political, rather than technical.

Numerous policy options are available in a *Policy Reform* scenario to torque energy development toward high efficiency and renewables. Fiscal mechanisms, such as carbon-trading systems, can be used to internalize environmental costs into energy prices. Carbon and other environmental taxes can be designed to be "revenue neutral" by offsetting other taxes, an approach already adopted in some European countries. Fossil-fuel subsidies can be eliminated. Progressively more stringent energy efficiency standards can be set by regulation. New financing initiatives and economic incentives can help spur investment in energy-efficient and renewable technologies. Increased research, development, and demonstration efforts can offer new technological opportunities. Market barriers to efficiency and renewables investment can be overcome through better information, capacity building, and institutional frameworks. Globally, initiatives to transfer technology and know-how can help make energy efficiency and renewables a foundation for the growing infrastructure of developing economies.

Food and land

The food–water–land nexus underscores the complex obstacles to sustainability. The simultaneous goals of providing resources for growing economies, meeting human needs and preserving a healthy environment are not easily met. Success will require the integrated perspective, political mobilization, and development philosophy of a *Policy Reform* scenario.

The *Market Forces* scenario offered a mixed picture for food and land. On the one hand, production keeps pace with steadily increasing requirements for agricultural commodities.

On the other hand, the doubling of agricultural output by 2050 contributes to unsustainable pressure on land and water resources. Increased chemical inputs to farming systems add to terrestrial and water pollution. Continued degradation of existing cropland amplifies the need for new agricultural land. Combined with losses to expanding settlement areas, forest areas and other ecosystems continue to be lost at the rates of recent years.

Global food demand increases in the *Policy Reform* are comparable to *Market Forces*. Indeed, the poverty reduction feature of the scenario implies somewhat higher average food consumption (Annex F-1). Animal products grow as a share of dietary intake in developing regions. Global requirements for cereals used directly as food increase by over 50 percent by 2050, from around 900 to 1,400 Mt. The demand for meat products nearly doubles, from 240 to 460 Mt.

The challenge of meeting global food requirements is greater in a *Policy Reform* context than in *Market Forces*. Stringent environmental goals limit land use, water availability, and chemical applications on farms. Forest and habitat protection reduce the scope for the expansion of agricultural land. Water sustainability constrains irrigation. Controlling water and land pollution and reducing land degradation require low-impact farming practices. A *Policy Reform* scenario would require a combination of measures to maintain farm yields and meet food demands while preserving ecosystems.

Sustainable farming

A transition to agricultural sustainability will require a "doubly green" revolution (Conway 1997). On the one hand, agricultural productivity must continue to improve in order to feed a larger and wealthier population. On the other hand, revised farming practices are needed that are compatible with environmental sustainability. The Green Revolution relied on new high-yield cultivars that require intensive use of chemical and water inputs. In the sustainability transition, the focus shifts toward complex farming systems, based on ecological principles, for enhancing productivity – knowledge-intensive rather than input-intensive agriculture (Swaminathan 1997).

In the *Policy Reform* scenario, a campaign for sustainable agriculture is launched. Market incentives, new regulations and subsidies stimulate the transition. Agricultural extension services and research and development efforts are reformulated to advance and disseminate innovations, many of which emerge locally. The gradual shift toward ecologically sound practices is supported by parallel efforts at poverty alleviation in the scenario, as increasing income allows farmers to diversify production and utilize advanced techniques (Scherr and Yadav 1996).

As fertilizer and pesticide applications decline, nutrient requirements of plants and pest control are met in other ways (Conway 1997). On farms, nutrient recycling, such as the capture of manure in combined crop-livestock systems and the use of large-scale composting, partially substitutes for fertilizer. Nitrogen-fixing plants are grown in rotation with, or in combination with, other crops. The development of pest-resistant plant strains and the widespread use of integrated pest management reduces pesticide requirements.

Biotechnology is an emerging tool that can complement conventional techniques for improving plant performance. It holds both great promise and significant risks. The capacity it brings for crop engineering can increase yields, reduce chemical inputs and improve the nutritional content of agricultural products. At the same time, it could decrease crop diversity if a small set of synthesized seeds become predominant; the fugitive genetic material from pest-resistant organisms could spread to off-farm ecosystems; and the world's farmers could become more dependent on transnational agribusinesses that control the technology (Raskin *et al*. 2002). In the last decade, the accelerated deployment of biotechnology in agriculture limited possibilities for prudent risk assessment and caused a loss of public confidence. A more deliberate and prudent approach, guided by the precautionary principle, would give biotechnology an important role in improving agricultural productivity while protecting the environment.

In the scenario, agricultural strategies in developing countries would aim to support sustainable livelihoods, benefit the poor and spare the environment. The program would be comprehensive – research, education and training of farmers, land reform, infrastructure development, support for non-farm rural enterprises and reform of economic policies. Programs for sustainable agriculture must be carried out in cooperation with farmers and be targeted to local needs (Leach 1995).

Fisheries and aquaculture

Growth in fish and seafood consumption in *Policy Reform* is comparable to *Market Forces*, reaching about 165 Mt in 2050 (Annex F-3). The goal of sustainable exploitation of marine and freshwater fisheries is achieved, in part, through better management, which increases the output of capture fisheries from 85 Mt per year currently to a sustained annual production of 100 Mt (FAO 1997c). Of the additional 15 Mt, about 10 Mt are due to the additional yield from replenished stocks and 5 Mt from reduced losses. The rebound in wild fish stocks follows the imposition of limited fishing rights and restricted access to overexploited global fisheries, while economic incentives, expanded markets and legal limits abate the current practice of discarding huge quantities of "by-catch."

Despite these improvements in production, some 60 Mt of global requirements cannot be met from capture fisheries on a sustained basis. Aquaculture expands to meet this additional demand, with production roughly doubling from the 30 Mt in the base year (FAO 1998a). In the past, aquaculture has led to pollution and pressures on freshwater resources (Brown *et al*. 1998). Sustainable fisheries policies reduced these through more careful site selection, conservation technologies, and pollution prevention.

Pasture and rangeland

The demand for livestock products increases in the scenario as populations and incomes rise. All else equal, this would drive up the requirements for pastureland proportionally. But changing farming practices breaks this equation. Modernizing farming systems relies on feedlots more than open grazing to expand production (Annex F-2). In addition, livestock productivity increases, especially in developing countries, as herd quality, disease control and grazing land productivity improve.

Livestock farming practices must change in order to reduce environmental pollution. For example, large feedlot facilities produce vast quantities of concentrated manure, which can cause nitrate pollution in groundwater and ammonia emissions, a precursor of acid rain. In the *Policy Reform* scenario, pollution from livestock wastes is moderated in part through greater use of integrated crop-livestock systems in which manure is recycled to crops as fertilizer (Durning and Brough 1991, Bender 1994). Sustainable practices are promoted through extension services, dissemination of indigenous innovations and pollution control regulation. In arid lands, where forage fluctuates with highly variable rainfall patterns, traditional forms of pastoral farming have been resilient in the past (de Haan *et al*. 1996).

However, under conditions of economic pressure to keep larger herds and restrictions on the movement of pastoralists, encroachment on marginal lands contributes to land degradation. In the scenario, policy processes negotiate a balance between the needs of pastoral farmers and protecting biomass resources.

Cropland, agricultural production, and trade

Global requirements for farm products doubles between 1995 and 2050 (Annex F-4). Additional requirements are met primarily through higher crop yields, which increase at rates roughly comparable to *Market Forces*. Increased regional production of food and feed in *Policy Reform*, therefore, leads to more cropped area. In OECD regions, cropland expands by 14 percent between 1995 and 2050 rather than the 7 percent of the *Market Forces* scenario (Annex F-6). Globally, cropland expands in *Policy Reform* only slightly more than in *Market Forces*, as regions with relatively low yields increasingly import from regions with higher yields. Of the nearly 200 million additional hectares brought into cultivation, over 70 percent is rainfed agriculture since expansion of irrigated farming is limited, particularly by the scenario goal of limiting freshwater abstractions to sustainable levels (Annex F-7).

Yields from irrigated farming systems tend to be significantly higher than rainfed systems. In 1995, global irrigated cereal yields were an estimated 70 percent higher on average than yields on rainfed land (Kemp-Benedict *et al.* 2002). As a consequence, the decrease in the fraction of new farmland that is irrigated – 75 percent between 1961 and 1994, compared to 20 percent in the scenario – accounts in part for the lower yield growth rates in the scenario relative to the past three decades.

Policy Reform initiatives to satisfy growing agriculture demands while diminishing stress on freshwater resources and preserving terrestrial ecosystems can succeed almost everywhere. This requires strong political will, new technologies and practices and a comprehensive set of effective policy instruments. This is necessary, but a number of arid nations would need to consider increasing their reliance on agriculture imports. By importing agricultural products, resource-scarce countries, in effect, import arable land and water, as well. The relaxation of long-held commitments to national food self-sufficiency would require a climate of geopolitical amity to assuage concerns about the security and conditionality of international food flows. It would also require rapid economic development and diversification to generate non-agricultural exports to pay for additional food imports.

The altered patterns of agriculture trade in *Policy Reform* are reported in Annex F-4. The level of food self-sufficiency declines in North Africa, the Middle East, and land-constrained countries in South Asia. Other regions – North America, Western Europe, Latin America, and the FSU – increase production in response to global markets. The trend in recent years in industrialized regions of withdrawing lands from production is reversed in the scenario in the context of heightened concern for maintenance of the world's ecosystems, a stable political climate and expanding economies in developing regions.

Land degradation

Land degradation in the *Policy Reform* scenario is greatly reduced relative to *Market Forces* trends. The preservation of land quality and ecosystem health is a cornerstone of the sustainability agenda. The pace of degradation slows over the next several decades with the adoption of sustainable farming practices and protection of marginal lands, after which land is restored rather than degraded. Cropland is lost at an average rate of 1.5 Mha per year to

2025, half the rate in the *Market Forces* scenario. From 2025 to 2050, previously degraded land is restored and returns to cropland, forest and other land types.

The changes in practices that reduce degradation vary with the character of the land and its use. Irrigated land that is vulnerable to waterlogging and salinization is spared by improved water conveyance and drainage systems, which also conserve water. Longer fallow periods reduce nutrient loss from shifting cultivation. Water-driven erosion is controlled by various farming techniques, such as hillside terracing and conservation tillage, that minimally disturb soil cover while improving soil water-retention. Agroforestry strategies combine crop and tree cultivation to stabilize terraces, reduce water run-off, fix nitrogen, and diversify farm commodities. Techniques such as these improve soil fertility, retain water, and reduce topsoil loss.

Built environment

The expansion of settlement areas is claiming arable land, forests and drylands at increasing rates. In the *Market Forces* scenario, the tendency toward greater spatial separation between home, work and commerce, and growing dependency on the automobile – the hallmark of post-Second World War suburbanization patterns in the United States and elsewhere – proceeds apace. Developing regions gradually converge toward this model as they develop. With economic growth, the extent of the built environment grows faster than population. Globally, an additional 320 Mha of forests and other land cover are converted for development over the next 50 years, or more than 6 Mha per year (Annex P-7).

In a *Policy Reform* world, the control of urban sprawl becomes a priority as part of the campaign to protect cropland, forests, and other ecosystems. Urban planning turns toward policies that promote more compact settlements. This effort is reinforced by parallel sustainability policies such as greenhouse gas emission abatement, that valorize ecosystems and temper automobile dependency. In the scenario, all regions gradually move toward settlement patterns currently found in Western European, rather than the more land-intensive style of North America (Annex P-7).

Forests and ecosystems

In merely 5 years, from 1990 to 1995, over 50 Mha of forest were lost, most of them in the developing regions (FAO 1999d). In the *Market Forces* scenario, the historic drivers of forest loss – conversion to farmland, pasture and settlements, and unsustainable forestry – continue unabated. In the push for near-term economic gain, the indirect and often non-monetarized costs of the liquidation of ecosystem assets do not figure in the calculus of investment and public policy. The lost services ecosystems provide – habitats, biodiversity, freshwater renewal, nutrient cycling, carbon storage, recreation, and esthetic value – stand as virtual costs to society and nature that are passed on to the future.

By contrast, in *Policy Reform*, the preservation of ecosystems becomes a key goal of development. Policies reflect the growing awareness of the vital role of forests and other ecosystems, both directly in the provision of economic goods and services, and indirectly in the maintenance of the biophysical basis for socio-ecological systems (Daily 1997). The long historic process of ecosystem loss gradually is halted. Special emphasis is placed on the preservation of old forests, wetlands, grasslands, and coastal zones.

Forest area is maintained and eventually restored through a diverse set of policies for protection and sustainable use. Subsidies and land tenure arrangements that favor forest removal are revised. The sustainable livelihood initiatives of the scenario expand opportunities for forestry by the poor, including the granting of secure land rights (Scherr and Yadav 1996). Finally, the sustainable management of forests by the timber trades becomes a priority everywhere. By 2025, forest areas have begun to increase (Annex P-7).

Summary

Numerous factors – economic growth, urban expansion, changing agricultural practices, and conservation policies – alter the use of land in the *Policy Reform* scenario. Land-use patterns are summarized in Figure 5.11, with more regional detail presented in Annex P-7.

Feeding a growing world population while maintaining the productivity of cropland, preserving habitats, and avoiding increases in pollution is one of the greatest challenges for sustainable development. There is no simple recipe, but rather it will require a combination of political will, a holistic vision, and a policy package tailored to local circumstance.

Policy Reform offers a picture of how this might be accomplished. Comprehensive governmental action builds capacity for research and extension services, provides adequate infrastructure, corrects perverse subsidies, institutes market incentives, and revises regulatory

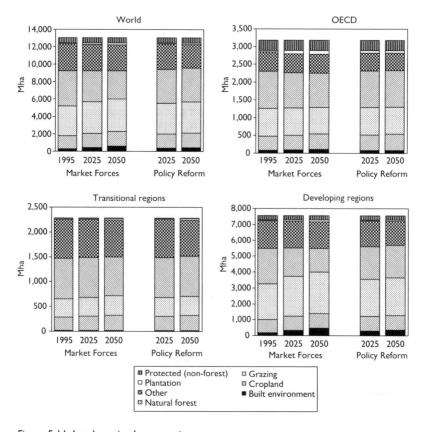

Figure 5.11 Land use in the scenarios.

frameworks. Gradually, agriculture, forestry, and land-use practices become more ecologically sustainable. The adequacy of water resources is critical to achieving this vision.

Freshwater

Abating freshwater degradation and scarcity is an exceedingly daunting task. The practical difficulty is reflected already in the sustainability goals introduced in Chapter 3. They accept that increased water stress is unavoidable, though at levels well below those in the *Market Forces* scenario. In *Policy Reform*, measures that moderate demand and enhance supply in order to meet the freshwater sustainability goals are introduced in each country.

Water sustainability is in some ways at cross-purposes with the poverty reduction and more equitable development aspects of the scenario, which increase water services to poor households and withdrawals for expanding developing country economies. Water requirements must be met far more efficiently and, to some extent, with nonconventional water sources. Certain arid countries, as we have seen, reduce water and land pressure by relying more on imports of agricultural products.

Efficiency improvement

Water has been used wantonly in the past. It has been undervalued, underpriced and underprotected. There is huge potential for using water more efficiently. Technologies and practices are available for reducing "water intensity," the water required per unit of economic service or end-use service. The dramatic decline in water intensity in the US manufacturing sector that was triggered by clean water legislation in the 1970s is shown in Figure 5.12 (Raskin *et al.* 1995). Through better housekeeping, water recycling and water-efficient manufacturing processes, the water required per unit of output was cut in half.

Changes in economy-wide water intensities in the scenarios are summarized in Figure 5.13, for the world and our three macro-regions. Aggregate water intensities decrease in both

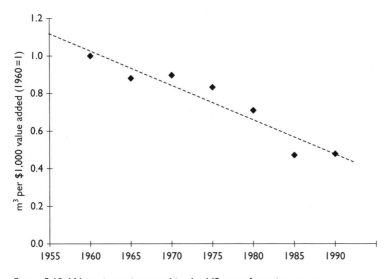

Figure 5.12 Water intensity trend in the US manufacturing sector.

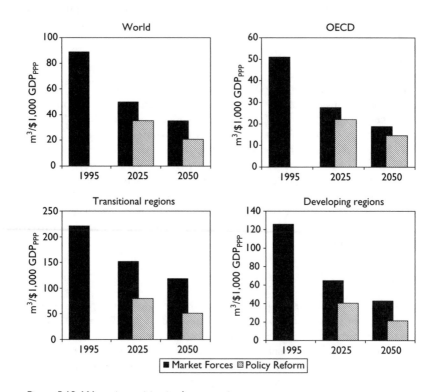

Figure 5.13 Water intensities in the scenarios.

scenarios as a result of two factors. First, the composition of economies shifts in the course of growth and modernization. As water-intensive sectors, especially agriculture but also heavy industry, contribute a diminishing share of aggregate economic output, aggregate intensities decrease.

Second, water is used more efficiently, as illustrated by the sectoral water intensity trends in Figure 5.14 for OECD regions. Other regions track these changes as they gradually converge toward industrial country practices in the course of development.

More efficient irrigation practices play an important role since irrigation withdrawals account for 70 percent of the global total. In the *Market Forces* scenario, water applications per unit of irrigated area increase, since higher-yielding crops in the scenario require more water. In *Policy Reform*, a variety of methods are deployed to lower irrigation water intensities (Postel 1992, Seckler *et al.* 1998). These include the substitution of conventional irrigation practices with more efficient spray systems, and covering open canals. In some areas, subirrigation is introduced, which raises groundwater to root level.

Opportunities to reduce water requirements are also seized in the household and service sector (referred to as "domestic" in Figure 5.14). Water-efficient appliances, tighter plumbing codes, and more sparing outdoor uses in water-short areas are promoted by fiscal and regulatory policies. Of course, domestic water intensities increase in developing countries with rising incomes, but far less than in *Market Forces*. They leapfrog the era of water inefficiency that high-income countries passed through to the early deployment of efficient practices.

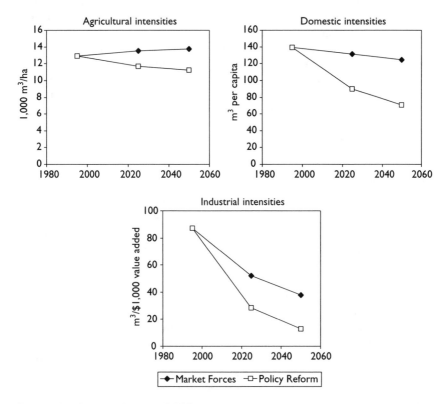

Figure 5.14 Water intensity in OECD regions by sector.

Water intensity in the manufacturing sector decreases in *Market Forces* as current trends toward greater efficiency play out in OECD regions, and developing and transitional regions converge toward these patterns. In the *Policy Reform* scenario, industry moves toward best-practice technologies, such as those currently used in Israel, where aggressive water-pricing schedules and standards for water-conserving technologies have led to highly efficient water use (Lonergan and Brooks 1994). Measures to reduce freshwater requirements include improved facility management, leak plugging, water recycling, opportunistic use of brackish water, and adoption of water-efficient processes.

Resource enhancement

In the past, water resource development focused heavily on the construction of large dams. Dams serve multiple purposes such as flood control, increasing reliable supplies, and hydropower electricity generation. Half of the world's 45,000 large dams were constructed primarily to meet expanding irrigation demands (WCD 2000). However, dam construction has been slowed by the exhaustion of appropriate sites, high costs and growing concern about the environmental impacts and human displacement.

In a *Policy Reform* world, these constraints and concerns moderate further dam expansion. Water resource enhancement relies on numerous niche opportunities. Innovation and

investment are stimulated by the upward adjustment of water prices to reflect true economic, social, and environmental costs. A program of technology development and outreach
promotes unconventional water projects. The environmentally sensitive expansion of
desalinization plants contributes substantially in coastal arid countries. Treated municipal
wastewater is recycled for use in certain agricultural settings. For example, by 2050 the
Middle East meets about 6 percent and North Africa about 10 percent of water requirements from desalinization and wastewater recycling compared to about 1 percent currently.
Some countries satisfy over 50 percent of their needs this way. Various low-cost and dispersed
techniques, including farm-level rainwater harvesting, conjunctive use (storing excess surface water in groundwater aquifers), and micro-dams, ease local water stress everywhere.

Water sustainability

The conventional paradigm is largely about water systems engineering, infrastructure
development and flood control. *Policy Reform* sees the emergence of a systems perspective
that respects the multi-faceted role that water plays in servicing human and ecological systems. The new paradigm elevates demand-side management through efficiency and adjust-
ing development strategies to conform to water constraints. It recognizes the importance of
stakeholder participation in fashioning strategies that balance the spectrum of legitimate
claims on water. Finally, it prioritizes the provision of clean drinking water and sanitary
services to all as a universal right.

This agenda plays out at the watershed and national levels, with the vigorous support of the
international development community. The momentum toward increasing water stress is
immense as illustrated by the *Market Forces* scenario. *Policy Reform* is able to moderate these
trends through efficiency, which leads to a decrease in global water demand rather than the
65 percent increase in *Market Forces*, and supply enhancement. In 2025, the economies of developing regions are 15 percent greater in *Policy Reform*, while aggregate water intensity is 40 percent lower. Regional patterns of water withdrawal are collected in Annex P-1. Improvements
in water use efficiency are reflected in moderated growth in all sectors (Annex P-2).

The *Policy Reform* scenario satisfies the sustainability goals set out in Chapter 3. Those
goals are modest. They recognize that the momentum toward greater water stress over the
coming decades can be moderated, but not eliminated (Figure 5.15). The basic indicator of

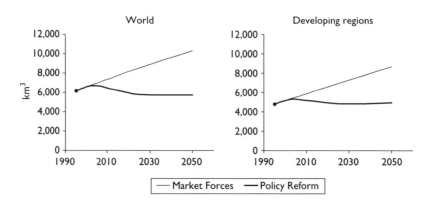

Figure 5.15 Water withdrawals in the two scenarios.

pressure on water resources, the "use-to-resource ratio," increases gradually in developing regions (Annex P-3). The fraction of global population living in areas of water stress grows from about 32 percent today to 36 percent by 2050. Because of population growth, the absolute number of people in water stress increases more rapidly than this, with much of the increase in the South Asia and Southeast Asia regions. However, in the *Policy Reform* scenario the incidence of severe stress (national use-to-resource ratios greater than 0.4) remains almost constant between 2025 and 2050, in contrast to the *Market Forces* scenario where it increases by about 30 percent over the same period. Meeting even these rather weak sustainability goals would require vigorous governmental initiatives.

Materials and waste

Each day a prodigious quantity of material flows into industrial economies and a waste stream of staggering complexity and volume flows out. It is found in gaseous, liquid, and solid forms of various toxicities. It accumulates as municipal garbage and discarded materials at mining operations, and discharges from agricultural and manufacturing operations. Huge quantities of synthetic chemicals are incorporated in products and dispersed into the environment as they wear.

The de-materialization of growing economies in *Policy Reform* rests on three strategic pillars – reduce, re-cycle, and substitute. First, the material content of final products and the material requirements of each unit of economic activity decrease rapidly. Wasteful packaging is sharply reduced, goods are re-designed to be leaner and last longer, and industrial processes become cleaner and far more materials efficient. Second, "waste" streams are re-conceptualized as resource streams that are thoroughly mined – products are refurbished and materials are recycled. A new "industrial ecology" treats the waste from one activity as a raw material for another. This model is built into the emerging infrastructure of developing countries. Third, the use of virgin materials and non-renewable resources gives way to renewable resources, such as the many forms of biomass.

Achieving these aims requires a comprehensive set of initiatives, with the details customized to local conditions. Key strategic levers are eco-taxes or trading systems to discourage virgin material use, elimination of perverse subsidies that encourage waste, and technology development. Conventional national accounting systems that treat environmental costs as economic goods, such as the clean-up costs for contaminated land and the health costs associated with pollution, are revised. Increasingly, producers are held accountable for the fate of their products; for example, through automobile take-back programs. Toxic and hazardous waste generation is radically reduced according to the schedule of the sustainability targets (Chapter 3).

The reduction of the throughput of materials, energy, and waste to lie within tolerable sustainability boundaries requires a thorough technological transformation. A moderation in consumerist lifestyles would ease the burden, but this would carry us beyond the premises of the *Policy Reform* scenario.

Chapter 6

Barbarization

Conventional Worlds scenarios posit that the potential environmental and social tensions induced by market-driven development can be resolved through the co-evolution of countervailing factors. *Market Forces* adaptations occur organically through the self-correcting logic of markets, institutional modernization, and human ingenuity. By contrast, *Policy Reform* assumes that governments at all levels make sustainability a political priority and that they proactively institute a comprehensive set of policy mechanisms for reaching long-range social and environmental goals.

A critical uncertainty is whether market and policy responses would be sufficient to cope with emerging socio-ecological stresses. There is no question that the feedback from unsustainable patterns of development will alter relative prices, accelerate technological innovation, and influence policy agendas in ways that counteract perilous trends. The issue is whether such adjustments will be too late or too little.

It may have always been the case that that human culture has changed more slowly than technologies and economies (Wright 2000). But now as technological and economic change accelerates on a global scale, such cultural lag could mean that institutions are able to respond only to yesterday's problems. Incremental market and policy adjustments could be overwhelmed by the de-stabilization and surprise that helter-skelter globalization has in the offing.

The environmental problems of several decades ago, such as air and water pollution, were local, clear-cut and immediate. By contrast, contemporary challenges, such as climate change and biodiversity loss, are global, highly complex and have long time delays. Similarly, the social problems of poverty and inequity are no longer locally bounded. They must now be addressed at a global scale where both the causal dynamics and potential remedies have grown diffuse and complex.

Rapidly changing global environmental and social challenges limit the scope for conventional responses at the national level – price internalization, regulation, subsidy, "trickle-down" economics, and social welfare programs. The autonomy of nations shrinks as their economies, cultures, and security become increasingly intermeshed in a maturing global system. Their future is linked to the common destiny: national sustainability is an aspect of a planetary project.

Globalism must complement nationalism. Indeed, the challenge of forging a global identity has antecedents in the long, and often violent, process of forming national identities over recent centuries. Indeed, that process continues today in some parts of the world. Now nations must find coherent global institutions for managing a global society, but over decades rather than centuries. Mindsets and ideologies that arose in a more parochial time may no longer be relevant.

The *Market Forces* reliance on automatic economic fixes is a matter of faith more than logic or experience. The *Policy Reform* vision of a series of corrections to harmonize market-driven globalization with sustainability principles, rather than reconsidering the fundamental premises of the model of development, is problematic. If market and policy responses are insufficient, the global development trajectory could veer toward a world of sharply declining physical amenity and erosion of the social and moral underpinnings of civilization. *Barbarization* scenarios consider the grim possibility that the march of conventional globalization is knocked off course by a general crisis.

We consider two variations of the *Barbarization* scenario – *Breakdown* and *Fortress World*. In the initial phases, the two variants tell the same story of increasing polarization, instability, and violence. But a host of factors – social conflict, economic disparity, environmental degradation, resource erosion, and institutional weakness – conspire to threaten economic welfare and civil order. The two scenarios bifurcate into separate trajectories based on the reaction of global institutions to the multidimensional crisis. These include transnational corporations, international organizations, and armed forces.

In *Breakdown*, the response of global institutions is inadequate and fragmented. The rising tide of chaos overwhelms all efforts to impose stability. In *Fortress World,* coherent and forceful action by global forces is able to contain the crisis militarily. An authoritarian world order is imposed, a kind of global apartheid that secures privilege and stability with police methods.

If global society is to navigate a transition to sustainability successfully, it cannot turn away from the perilous shoals that may lie along the way. *Barbarization* scenarios are warnings that help reveal dangerous patterns that can increase the readiness to act to counteract the conditions that could initiate them. A full examination of the social scapes of the future must consider all plausible possibilities in order to illuminate, motivate, and guide actions today.

The scenarios unfold

The point of departure of *Barbarization* scenarios is the set of driving forces that initially propels all global scenarios. As we have seen, these include economic and political changes, persistent poverty and deepening inequity, growing and more urban populations, environmental degradation and technological innovation.

The story begins in the early years of the twenty-first century. A kind of market euphoria had gripped the developed world in the last decade of the twentieth-century. But that is shattered by economic and political crisis. A global economic recession cools the optimism of the 1990s. The 11 September 2001 terrorist attacks on the United States and their aftermath alter geopolitical conditions and elevate security concerns in international affairs. The world is reminded that globalization is not a smooth ride to prosperity and peace. It is a rocky path full of twists and turns.

Contrary to academic pronouncements of "the end of history," the future seems contested and full of uncertainty. Deep fissures across the cultural landscape are revealed. During these years, the powerful nations of the world have an opportunity to lead a process of policy reform to create a form of globalization that is more inclusive and sustainable. But the moment is squandered. A cycle of violence and retribution exacerbates anger, suspicion, and polarization. Eventually, a coordinated international campaign controls world terrorism, but episodic attacks renew fear, invigorate police vigilance and erode compassion and civil liberties.

The era of sustainability was marked by a decade of international conferences, beginning with the Rio Summit of 1992 and ending with the World Summit on Sustainable Development of 2002 in Johannesburg. High-profile global assemblies addressed a host of social and environmental topics, but produced more high-minded rhetoric than substantive change. After Johannesburg, political momentum evaporates and the voices for a sustainable alternative fade. Environmental and social non-governmental organizations lose their influence and begin to atrophy.

The thinking in policy circles centers on a so-called "new realism" that advocates limited global market development with the support of a strong security apparatus. It abandons the market utopianism of the 1990s and the inclusive dream that global economic growth could enrich and modernize all. There are those who argue, drawing on the lessons of history and science, that the drift of development is not realistic at all, but, rather, endangers civilization and the planet.

These isolated voices of dissent are not suppressed – they are simply ignored as irrelevant anachronisms. A younger generation comes of age. Raised in the interconnected world of the twenty-first century, they find the new global realism both natural and inevitable. To them, the graybeards' occasional paeans to civic life, local culture, and national loyalty are a quaint artifact of an earlier phase of history, like the Greek city-state. Acquiescence reigns where once there was hope for a global movement for ecology and justice. The exceptions are terrorism and the desperate vandalism of anti-globalization anarchists, who are violently suppressed.

The injunction of the twentieth century sustainability movement to take responsibility for the well-being of future generations fades. Time horizons contract to the next quarter's profits, the hottest technological gadgets and the latest chapter in the war against terrorism. The focus of most lives in rich countries and affluent strata in developing countries is on feathering their own nests. The media reinforces a philosophy of self-interest and immediate gratification. Eschewing quaint old notions of the family of man, personal identification and loyalty shrink to a small circle of family and friends. Meanwhile, the impoverished billions grow more desperate and bitter.

The new ideological climate spawns a revised role for government. As the legitimate guarantor of stability, it establishes all the institutions necessary for peaceful world commerce. At the same time, government shrinks elsewhere as the process begun in the twentieth-century of deregulation and privatization of its functions proceeds apace. With the waning of civic commitment, the vestiges of the welfare state, including social services and safety nets for the poor, erode. Development aid is a victim of these changes. In the early twenty-first century, United Nations' targets for foreign assistance from high-income countries were already ignored. Over the following decades development assistance vanishes, except for disaster relief and crisis management.

The control of states over their national economies diminishes as they increasingly react to the imperatives of international financial and capital markets and the dictates of economic globalization. Management of the global economy is ceded to new formations representing transnational interests such as the World Trade Organization, multinational banks and multinational corporations. The new power alignments are ascendant, dominating the residual forces of nationalism, protectionism, and isolationism. For the most part, this evolves through peaceful assimilation into the hegemonic logic of globalism. However, some religious extremists, terrorists, and local tyrants cannot be placated by the promise of participation in the global emporium. They become casualties of an unending war on terrorism.

The world population begins to split into two huge sectors. The globally connected "haves" account for 20 percent of the world's population but 90 percent of its wealth. They are found in all corners of the world, in the growing pockets of affluence in poorer countries and throughout rich countries. Overlaid on this primary economy are the billions of "have-nots," politically marginalized and excluded from the modern global market, except as a source of cheap labor. Both absolute poverty and the level of inequality increase.

Scientific and technical research continues to move from the public to the private domain. Research is geared toward information technology, communications, and new commodities for the global consumer market. A relatively few large corporations concentrated in rich countries and enclaves capture the vast profits from innovation, reinforcing global inequity. Meanwhile, research languishes on such fundamental problems as the environment, global change, and public health.

While lacking economic opportunity, the poor do have abundant access to global communications and entertainment media. They are tantalized by images of opulence and dreams of affluence. Their increased exposure to affluent travelers accentuates the immense differences in lifestyles between rich and poor. Intellectuals in poor regions become especially embittered by the inequity of world development and the denial of opportunity. They feel cheated, that the wealthy have pre-empted their options. A new social actor emerges – educated, downwardly mobile, and angry – to fan the flames of discontent.

The demographic transition begins to reverse among the poor as birth and death rates both rise. Populations soar, contributing to the cycle of poverty. A huge international youth culture, connected by common styles, fads and attitudes, takes shape. Numbering in the billions, the new "global teenager" is consumerist and nihilist, tendencies forged by entertainment and advertising imagery that reach every corner of the earth through the expanding information revolution (Schwartz 1991).

But the enticing visions that pour from TVs and cyber cafes are unattainable. The increase in expectations is on a collision course with the decrease in access. Global youth lose their moorings, and social norms lose their control. Many turn to drugs and violence. Others join the armies of migration, much of it illegal. The desperate and displaced, lured by the promise of prosperity, move in waves toward the enclaves of the rich. The affluent respond with growing xenophobia as the poison of social polarization deepens.

Despite some successes in abating pollution in high-income areas, general environmental conditions deteriorate. Both parts of the dual world economy contribute. The modern global economy increases greenhouse gas emissions, claims on resources, and chemical pollution. Efficiency of resource use improves, but not fast enough to outpace the surge in economic growth. The other economy of the overpopulated world of the poor is also harsh on the environment. Anarchic urbanization and expansion onto marginal rural lands destroy ecosystems and deplete resources. Unsustainable agricultural practices accelerate soil degradation and deforestation. The thirst for scarce freshwater leads to conflict along shared river basins.

Brittle marine fisheries collapse under additional pressure, depriving a billion people of their primary source of protein. As the ecosystems of the world are lost, degraded and fragmented, species extinction accelerates and the world's biological and genetic endowment is impoverished. Climate change exacerbates environmental and resource problems. Famine in Africa and elsewhere grows more frequent and more severe, while the response capacity of relief agencies slackens. Public health suffers from the emergence of new and the resurgence of old diseases.

Conflict begins to flare in hotspots of social discontent, environmental crisis, and economic disruption. Discord amplifies and spreads, flamed by regional disparities and competition. Old ethnic, religious and nationalist fissures fuel an outbreak of armed conflict and violence. Poor countries begin to fragment as civil order collapses and various forms of criminal anarchy fill the vacuum. International investment in troubled regions plummets.

The unfolding crisis takes its toll on the economies in prosperous areas, as well. Public and private resources are diverted to security. Environmental protection and infrastructure are neglected, hastening degradation and decay. The affluent feel the sting as their economies falter and standards of living fall. Increasingly, the rich perceive the poor multitudes as a threat to their well-being, and even survival. Constant reports of migration, crime, terrorism, disease, and global ecocide feed the sense of peril.

Eventually, the global economy sputters and international institutions weaken. The retreat of globalization is particularly devastating for industrial economies highly dependent on trade and imported natural resources. With the failure of markets and investment, technological progress halts and the capacity to maintain existing equipment wanes. Decades of development are reversed. The global crisis spins out of control.

An era of regional fragmentation, civil disorder and a poisonous intolerance descends. Some countries devolve to a jagged-glass pattern of city-states, shanty sprawls, and nebulous regional formations. Some formerly industrial countries join the ranks of impoverished states. Incessant conflict, chaos and uncertainty foster pervasive fear and hopelessness. Economic development ceases and technological progress stagnates (except for better security devices for the privileged). No individual country is able to lead or control the others.

In this crisis atmosphere, the remaining institutions of global authority come together to consider emergency action. The organs of transnational corporations, the military representatives and international governance form the so-called Coalition for the Future. It issues a manifesto declaring a global emergency and a program of action, the "mobilization for peace and rectification." It announces the "temporary" imposition of martial order featuring coordinated interventions to protect natural resources, police the movement of people and keep the peace. Global military forces are placed on alert.

The *Breakdown* variant

But the Coalition's ambitious mission is undercut by conflict within its ranks. The years of competition and rivalry are a fragile basis for a globally coordinated response of such complexity. Mired in power struggles and mutual recrimination, the grand program for planetary salvation can muster only fragmented and ineffectual interventions.

Meanwhile, the vicious cycle of chaos, conflict, and desperation spirals forward. The security apparatus cannot contain the swirling tide of violence from terrorism, organized crime and political extremists. Civil order collapses in swaths across troubled regions. Bands of irregular armies and criminal groups struggle for turf and control, terrorizing people and causing great damage. Refugees fleeing from chaotic zones destabilize neighboring areas, inadvertently contributing to widening waves of disorder. Anarchy is on the march.

The areas where government still functions mobilize to stem migration and terror, but draconian police powers are not enough. Global economic, finance, and governance systems continue their collapse, though the media lingers to spread fresh news of upheaval. Civil disorder reaches even the rich countries.

Eventually, social, cultural, and political institutions disintegrate in this cascading upheaval. The world de-industrializes, technological capacity regresses, and agricultural systems collapse. Public health deteriorates, death rates surge, and global population contracts. Twentieth-century dreams of a prosperous and sustainable global future have turned into the nightmare of a broken world of primitive communities and roving tribes.

It is a bitter irony that *Breakdown* tempers two factors that contribute to it – social inequity and environmental degradation. Equity increases, but only because everybody gets poorer. The environment slowly heals, but only because the world economy has collapsed. These fractured and impoverished conditions persist for many decades before social evolution again becomes possible.

The *Fortress World* variant

The Coalition for the Future is fraught with internal friction at first. But comprehending the risk of a global breakdown, the various global actors join and are able to overcome their antagonism. They mount an effective campaign to stem the global crisis and protect their common interests. A lasting global alliance that provides the organizational basis for a new world order is forged.

The global movement for the imposition of coherence and control is strongly supported by the world's rich and elite groups. Many are uneasy about the authoritarianism of the Coalition. They see it as a regrettable necessity, the only alternative to the utter collapse of wealth, resources, and civil order. There is significant organized opposition in the under-developed regions, but it is politically and militarily fragmented. Failing to command the active support of the impoverished masses, the resistance is no match for the sophisticated military of the global forces. Even among the less privileged, most people are weary of the chaos and have little appetite for resisting the forces of order.

The elite retreat to protected enclaves in rich nations and to strongholds in poor nations. The majority outside the fortresses are mired in poverty, their movements monitored and their rights restricted. Draconian police measures control social unrest, migration, and spo-radic episodes of armed resistance. Strategic mineral reserves, fresh water, and important biological resources are put under military control. Environmental edicts that protect the global commons of air and ocean resources are issued and repressively enforced.

In the fortresses, technological culture is preserved and the process of innovation continues. Pollution within the fortress is radically reduced through obligatory use of clean techno-logy and solar energy. Residual pollution is transferred to deposition sites in the so-called "open areas," the vast regions outside the privileged enclaves. Natural resources are extracted from there for use in the modern economy that functions across the globally linked archi-pelago of fortresses. Some of the open areas are well preserved; the resorts, hunting grounds, ecological zones, and playgrounds of the affluent.

The elites have halted the advance of barbarism at their heavily guarded gates, but only by consigning the majority of the human race to destitution and oppression. They are entrenched in bubbles of privilege amidst oceans of misery, descendants of the "gated cities" and personal security industry that first arose in the late twentieth century. For those unfor-tunate enough to be born poor, life is Hobbesian – nasty, brutish, and short. The long-term stability of this system of global apartheid depends on the capacity of the society of enclaves to maintain control over the disenfranchised. New generations will grow restive and angry. A general uprising of the excluded could threaten continuity. The *Fortress World* may contain the seeds of its own destruction. But it could last for decades and, then, what next?

Chapter 7

Great transitions

Conventional Worlds scenarios offer visions of gradual adjustment and essential continuity of future values and institutions with those of the industrial era. *Barbarization* scenarios glimpse the unhappy possibility that conventional development fails to cope with the environmental and social stresses it induces, and civilization unravels. We turn now to a third path, *Great Transitions* scenarios in which global society, rather than descending into cruelty and chaos, evolves to a new stage.

Beyond sustainability

A transition to a sustainable future would need strong economic and technological foundations. The transition requires sufficient material wealth to provide a decent standard of living for a growing population. It must deploy appropriate technology for harmonizing increasing human activity with environmental preservation and the sharp reduction of absolute poverty. The emphasis on economic growth is the defining characteristic of the *Market Forces* scenario. The *Policy Reform* scenario adds these technological and social dimensions.

To the degree the reform agenda succeeds, the dangerous drift toward environmental degradation and social polarization would be ameliorated. The historical challenge is immense, but the policy tools and technology are already available for tilting development toward sustainability goals. The primary barrier is insufficient political leadership for grafting such a comprehensive set of corrections onto the market globalization paradigm. It would require extraordinary vision from the world's political leaders. It is difficult to visualize how the necessary political will would emerge within a world economic system and polity that are conditioned by the conventional development paradigm.

Substantial pitfalls and uncertainties lie along the *Policy Reform* path. In addition, a profound normative question can be asked: is this path desirable? *Policy reform* tames the environmental and social predations of the market-driven development model, but it is an engineered world with little leeway to cope with unexpected changes. It adjusts the system without transforming the ultimate drivers that shape it – values, understanding, and lifestyles.

Great Transitions scenarios add a new feature – a values-led shift toward an alternative global development vision. Materialism, consumerism, and individualism are tempered by the greater valorization of more qualitative desiderata, such as spiritual, cultural, and intellectual fulfillment, quality of community and enjoyment of nature. The new values are rooted in an ethic that recognizes the connectedness of humans to one another, to the wider

community of life, and to the future (ECI 2000). Fundamental human values are empha-
sized – fairness, freedom, compassion, tolerance, honesty, and solidarity. The new cultural
context opens fresh opportunities for ameliorating conflict, eradicating poverty, and building
social cohesion. Revised lifestyles and strong environmental values, along with better
technology, diminish the human footprint on the planet.

Market Forces, *Policy Reform*, and *Great Transitions* embody very different strategies as
illustrated in Figure 7.1. In *Market Forces*, human well-being is correlated to consumption.

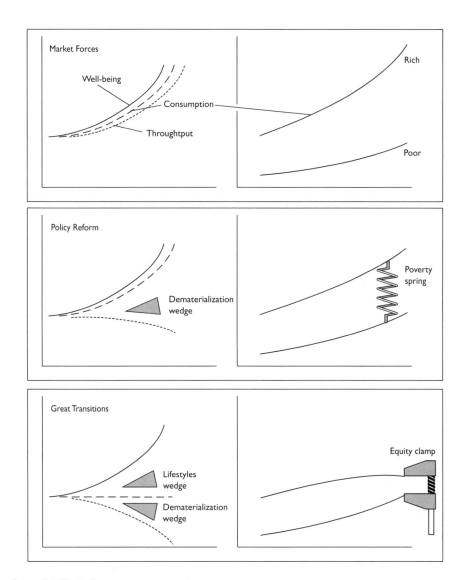

Figure 7.1 Tools for a transition.

Source: Raskin *et al.* (2001).

In turn, rising consumption drives parallel increases in the use of natural resources and environmental impact ("throughput"). *Policy Reform* maintains the connection between well-being and consumption. However, both become decoupled from throughput by the "dematerialization wedge" – the deployment of resource-efficient technologies and renewable sources of energy and materials. *Great Transitions* adds the "lifestyle wedge" that breaks the link between consumption and well-being, as nonmaterial fulfillment plays a greater role in defining human aspirations.

Total environmental impacts are the product of two factors, the scale of human activity and the impact per activity. The dematerialization strategies of *Policy Reform* decrease impacts per activity by inducing high resource efficiency and the use of clean and renewable technologies. The changes in lifestyle and patterns of consumption in *Great Transitions* have the additional effect of lowering activity levels.

The three scenarios differ in their approach to social equity, as well, as suggested by the right panels of Figure 7.1. All of the scenarios foster greater equity to various degrees. In *Market Forces*, the economies of developing countries grow more rapidly than high-income regions as globalization leads to gradual international convergence. But the process is extremely slow. Because contemporary inequities are so great, the absolute difference between rich and poor countries continues to widen for many decades. Moreover, inequity within countries tends to increase as those able to participate in the modern global economy prosper while the poor are left behind.

In *Policy Reform*, governments take action to substantially reduce absolute destitution through a set of strategies targeted to provide sustenance and livelihoods for the poor (the "poverty spring"). While helping those at the bottom of the economic pyramid, global and national inequity is only partially abated. In *Great Transitions*, a strong sense of global and community solidarity fosters far more egalitarian societies (the "equity clamp").

Policy Reform rejects a world of environmental unsustainability and rampant poverty. *Great Transitions* rejects a culture in which the endless accumulation of things is the path to fulfillment. The quest for material extravagance comes to be seen as an anachronism that impoverishes the intangible sources of wealth in culture, creativity, and human relationships. Interest turns to cultivating a sustainable relationship with nature, the resurrection of community, and the search for meaning.

This constellation of values is alive today as a minority and oppositional outlook. It may portend a worldview for a future day if the ferment grows and coalesces into a broad consensus on the general direction for an alternative development paradigm. Underlying the inevitable range of views on how to address specific issues, the new paradigm would rest on three pillars for a new civilization: ecological sustainability, human well-being, and unity in diversity.

Ecological sustainability means limiting consumption, waste generation, and population as an existential precondition for the flowering of human civilization and life on the planet. Human well-being has quantitative and qualitative dimensions – an economic system that provides a comfortable standard of living to all, on the one hand, and the space to pursue psychological, cultural, and spiritual fulfillment, on the other. The well-lived life is about the quality and authenticity of experience – creativity in different realms, the search for truth and beauty, human solidarity, and a harmonious relationship to nature.

The third pillar, the principle of "unity in diversity," requires governance arrangements that balance global goals with pluralism in meeting those goals. The collective interests in controlling climate change, preserving ecosystems, maintaining world peace, and securing human rights place limits and conditions on regions and communities. While conforming to them is an obligation to the common good, the means are decentralized and diverse. Governance systems must have the capacity to respond adaptively to surprise and possibility. Decentralized, cooperating networks are better at this than rigid top-down structures. The right to explore cultural alternatives is fundamental for resilience and freedom in a unified and heterogeneous global society.

Social actors and the seeds of change

Great Transitions envisions a profound change in the character of civilization in response to planetary challenges. Such transitions have happened before at critical moments in history, such as the rise of great cities thousands of years ago and the emergence of the modern era over the last millennium. All components of culture are transformed in the context of such a holistic shift – values and knowledge, demography and social relations, economic and governance institutions, and technology and the environment. These dimensions of transition reinforce and amplify one another in an accelerating process of change.

Intimations of such change are already present. Though inchoate and undeveloped, these are manifested in the engagement of civil society in the development discourse, in expressions of alternative values, and in the actions of some sectors of government and business. The possibilities for redirection could expand as the dynamic of planetary change proceeds and crises deepen, and if the lure of a new global vision seizes the popular imagination. But it takes a long view to begin to imagine the actors and processes that might usher in a *Great Transition*.

The second half of the twentieth century saw the emergence of three important social actors on the global stage – intergovernmental organizations, transnational corporations, and nongovernmental organizations (NGOs). The UN was forged by the horror of twentieth-century world wars and the dream of world peace and understanding. With its constellation of specialized agencies and associated international banks, it is a key unit of international governance. But despite its many achievements, the early hopes that the UN could establish an era of human rights, amity, and development have not been fulfilled. The original vision of a global civil service with a broad mandate was compromised by administrative inefficiency, funding insufficiency, and the absence of any real power. The organization was too often used by its members as a forum for national agendas and political manipulation, especially during the Cold War period. Yet it represents the legitimate voice of the governments of the planet and will have a critical role in shaping the global future. Other supranational groups – OECD, Group of 77 (Southern Group), European Union, regional development banks, and new regional formations as they arise in the future – will bring their own mandates and viewpoints to the table.

The significance and power of transnational corporations have been rising since the Second World War. These huge business combines command immense human, financial, and natural resources. They have become an economic force with a truly planetary reach and a growing ability to influence world markets. As capitalist entities, they are competitive, expansionist and driven by the bottom line of profit. But their concern for the long-range stability of the global economic system and their sensitivity to consumer perceptions open opportunities for accountability and sustainable business practices.

Nongovernmental organizations are more recent to explode on the global scene. These voices of civil society speak through groups of astounding diversity in size and purpose. They include both large globally connected organizations and a legion of others dispersed at the grass roots level. The rise of thousands of NGOs in support of a host of environmental, social justice, and labor issues has been one of the most dramatic processes of recent decades. For many concerned about democracy and sustainability, it is a hopeful development. NGOs play an important role in building public awareness, whistle blowing on corporate misdeeds and participating in global, national, and local policy negotiations.

The evolution and behavior of these sets of actors – international organizations, multi-national corporations, and NGOs – will be critical in shaping global development. In a *Great Transition*, one must imagine them coevolving, reflecting, and reinforcing new mandates and values. The United Nations becomes an effective global governance forum able to advance an agenda that balances growth with equity, human rights and environmental sustainability. The corporate world energetically works with government and non-government organizations to implement transparent and comprehensive codes of conduct for socially and environmentally responsible practices. NGOs increase their reach and sophistication, and move from fragmented issue-oriented politics to a more unified concern with the interconnections between the various problems they address.

But what could lead to these new roles for social actors? The prospects depend on a wider reconsideration of the development model. The necessary condition is increased public awareness of global risks, and a willingness to act. Certainly a greater appreciation of the linkage between issues would help foster the unity of mind-sets and movements that might otherwise remain fragmented and ineffectual. Science can contribute a sense of urgency and legitimacy, as it better understands the potential for shifts in the global ecology, threats to life-support systems and risks to public health. The conditions for change would come to pass, if at all, through a process of slow change and acute events. Awareness of perils and opportunities mounts gradually in a process punctuated by crises that can stimulate rapid changes in consciousness.

A severe climate event or a war can cut through layers of denial as people search for underlying reasons and causes. Consider a hypothetical drought of the future that ravages both Southern Europe and North Africa. The contrast in peoples' fates – aid and comfort for Europe and chaos and death for Africa – is a vivid symbol of inequity. Deep resentment boils into anger throughout the developing world. It triggers an escalating battle between the desperate and the affluent, as political and religious armies of the poor attack the icons and forces of globalization. Unexpectedly, the emergency feeds into changes in public awareness that had been brewing for years. It galvanizes masses of people into coordinated action around the world. They demand coexistence, a more equitable global society, climate stabilization, and sustainability.

A real-world example comes from the 11 September 2001 terrorist attacks, a decisive event that defines a "before" and "after" in the flow of historic time. The events as they unfold over the coming years may reinforce polarization and the forces of *Barbarization*. But the greater awareness of the syndrome of poverty, humiliation, and anger could also prompt a politics of constructive international engagement and inclusive globalization. The response to new surprises in the years to come will depend on the depth and character of awareness and values – and visions for the future.

Narrowing the focus

The *Great Transition* scenarios may seem idealistic and improbable from the perspective of today's values, conflicts, and trends. But they are possible and may even be necessary to achieve the goals of sustainability and equity. They imply structural shifts in the global trajectory at some point in the future that go beyond incremental changes. They could emerge, if at all, either through a fundamental departure from the current path, or as a new start following a destructive period of barbarism.

The second possibility – that a good society emerges from the ashes of desolation – may be intriguing as a matter for contemplation and creative fiction. But it is not instructive for policy-making or for formulating actions to avoid *Barbarization*. That is, unless barbarism was thought to be inevitable, in which case strategies might be considered for planting the seeds for renewal sometime in the future and protecting them during the coming dark ages. Or if one believed, as some extremists and terrorists apparently do, that the only path to a better world must pass through chaos and anarchy, the perverse conclusion might be drawn that this process should be accelerated. It is our conviction, however, that *Barbarization* is avoidable, so such extreme pessimism and ignominious ethics, while not inconsistent, will concern us no further.

No single blueprint can capture the immense range of possible *Great Transition* scenarios. Reasonable people can disagree on the desirability, stability, and likelihood of alternative visions of a golden age in this century. Our taxonomy of idealized scenarios (Chapter 2) introduced two variants to reflect at least some of this diversity: *Eco-communalism* and the *New Sustainability Paradigm*.

Eco-communalism envisions a patchwork of semiautonomous and largely self-reliant communities. With high equity, a strong environmental emphasis and low populations, such a world would be quite sustainable in many ways. However, its resilience would be subject to the threat of aggression. Conceivably, a culture of peace and greater understanding of human behavior could sharply reduce such risks. But in the absence of a strong commitment to global governance, the largely self-governing and independent communities, perhaps in the sway of a charismatic tyrant, could become belligerent and colonial.

A fundamental problem with *Eco-communalism* as a global system is identifying a plausible trajectory that gets there from the present situation. The developing connectedness and complexity of the world economy suggests that any transition to such a society would be mediated through a series of other social formations. There appear to be two possibilities. First, *Eco-communalism* might evolve from a *New Sustainability Paradigm* world. If a powerful consensus arose for localism and autonomy, interregional linkages might whither, especially if advances in small-scale technology reduce the need for global infrastructure and trade. Second, *Eco-communalism* might emerge in the recovery from *Breakdown*. Under conditions of reduced population and the rupture of modern institutions, the rise of a world of communities guided by a "small-is-beautiful" philosophy is conceivable. In either case, our focus must turn to the *New Sustainability Paradigm*, as a precursor of other *Great Transition* formations and an alternative to *Barbarization*.

The *Great Transition*: a brief history

During the whole of its history, the human species struggled to survive and expand its dominion. Material accumulation was adaptive in the sense that food reserves, land, and gold gave individuals, groups, and societies a competitive advantage. Eventually, market economies came to dominate world development, bringing new incentives for innovation and expansion, and the values of competition, individualism, and consumerism. By the twenty-first century, the system based on expanding consumption and production had reached a global scale and become ecologically problematic.

It is clear that during the first decades of the new millennium an extraordinary shift in human history began to take shape. It did not seem that way at first. But looking back from the future with the benefit of hindsight, beneath the surface, incipient forces for change were brewing. The revised development agenda came in three periods – crisis, reform, and transformation. These correspond roughly to what some called at the time *Market Forces, Policy Reform,* and *New Sustainability Paradigm.* In truth, the long view of history helps us to see that these are actually three phases of a unitary process of change – the *Great Transition* (Raskin *et al.* 2002).

In the first phase, the policies and ideology of corporate globalization seem destined to define the character of world development for the indefinite future. Global environmental degradation moves into fast-forward mode, social polarization deepens, and economic instability mounts. Then, in the first few years of the twenty-first century, a global economic slowdown and the jolt of world terrorist attacks on the United States stimulate a vigorous debate on globalization and new global alliances.

The lesson ultimately drawn by world leaders is that the march to an open world economy must be accelerated and extended to the far corners of the underdeveloped world. Economic growth and the universal spread of market-enabling institutions is understood as the precondition for world stability, poverty reduction, and environmental preservation. The "growth first" philosophy is pursued with single-minded intensity. Global governance succeeds in the sense of facilitating trade liberalization and capital mobility, and maintaining peace and world order. But the international response to environmental and social issues is either totally inadequate, as in the case of the Kyoto Protocol on climate change, or nothing more than high-minded rhetoric, as in the constant appeals for poverty reduction.

Over the next decade, signals of ecological instability, biological destruction, and environmental links to human health become more frequent and more urgent. It is difficult to turn on a television or read a newspaper without finding new and alarming reports. The concern mounts that some environmental problems could be approaching thresholds where extreme events could kick in – severe climate events, ecosystem collapse, and health pandemics. Advances in the science of large-scale planetary shifts confirm the rationality of these fears. Many begin to see the risk of ecological calamity as an unacceptable cost of heedless economic growth.

Meanwhile, the global poor, the billions left behind by economic globalization, grow more desperate. Their plight increasingly is expressed in migration pressure, civil disorder, and political dissent. As the hopes for better living conditions of masses of people are lost, the rising tide of resentment and anger poses clear threats to security. Concern about the threat of social unrest and conflict rises everywhere.

Many come to the conviction that governance has grown too weak and that the world has become too reliant on the profit motive as the primary engine for development. The general public is apprehensive and restive, as NGOs, and eventually the mass media, build awareness of the unfolding global crisis. Concerned for the integrity of the international economic system, the ranks of powerful corporations calling for action swell.

As the clamor for leadership intensifies, a new generation of politicians comes forth to launch a process of *Policy Reform*. In this second phase of transition, hard-hitting sustainability goals are set at all levels, and an unprecedented host of new initiatives are enacted to redirect development toward sustainability. Its initial accomplishments are impressive and for a period a gentler form of globalization seems feasible. But *Policy Reform* fails to maintain its momentum, as political will waxes and wanes with the economic calculus of powerful interests and countries. The policy initiatives are unable to counteract the underlying market dynamics driving inequitable and unsustainable growth.

The partial reforms further stimulate the movement for more fundamental changes and a new global bargain. People's motives are complex, with both pragmatic and idealistic aspects. They are pushed forward by renewed fears of cataclysmic global crises. They are pulled forward by the lure of a better planetary culture. The necessity for change has a powerful partner – the desire for a "new sustainability paradigm."

With a growing disenchantment with consumerism as an organizing principle for their lives and communities, people seek positive alternatives. They explore more fulfilling and ethical ways of life that offer a stronger sense of meaning and purpose. In the wealthier areas, the values of simplicity, tranquility, and community begin to displace those of materialism, competition, and individualism. Many people opt to reduce their work hours and income to free time for study, art, hobbies, relationships, and spiritual search. The emergence of these profound personal, philosophical, and cultural dimensions marks the opening of the third phase of the *Great Transition*.

Parallel with these value shifts, and interacting with them, immense changes are underway in the developing world. A cultural renaissance, rooted in pride and respect for tradition, and an appreciation of local, human, and natural resources, unleashes a new sense of possibility and optimism. A young generation of thinkers, leaders, and activists catalyzes a fresh debate within their countries on the meaning of development and strategies for success. They join and enrich the global dialog, giving the emerging movement a truly planetary character. A rising popular consensus in North and South, East and West, declares conventional development wisdom insufficient and undesirable.

The new values seem to rise spontaneously from a thousand sources – spiritual communities, political movements, traditional cultures, and senior citizens. Some turn toward esoteric sects, but they are the minority. Youth from all regions and cultures drive the process of change as they rediscover idealism and a new collective identity. Gradually, a worldwide ferment coalesces, with untold millions joining the search for new forms of social existence. A global discussion seeks a core of shared goals for survival and cultural renewal. The exchange of ideas and sense of global unity builds on networks that span the planet, linking individuals, groups, organizations, and governments. The Internet is the natural medium for the new consciousness, providing a sense of immediacy and connection to a diverse and pluralistic movement – and powerful channels for democratic participation.

Global meetings and festivals explore and celebrate the rising movement for a new sustainability paradigm. Civil groups organize international political movements, and governments and corporations alike pay close attention. Eventually, some communities opt for alternative lifestyles and economic practices. Some stress high-technology solutions, some prefer frugality, and some practice "small-is-beautiful" utopian visions with an eco-mystical relationship to nature.

Conventional interests react against the new consciousness, hoping to slow down and reverse the energy for change. Some form a self-styled Coalition for the Future that presses for an authoritarian solution to address what they see as a triple crisis of: the environment, underdevelopment and the movement for a *Great Transition*. Their main impact is to raise the fear of a *Fortress World*. Ironically, this galvanizes masses of people into action to preserve the momentum for a new development paradigm. As the tension mounts, the Coalition fades into the mists of history. The tide of change rises inexorably.

The planetary movement gradually gains the allegiance of the majority of people in many places and, through its representatives, the levers of governance. Under popular pressure, governments, global NGOs and corporations negotiate a New Planetary Deal. It has three major elements: an Economic Bargain, a Knowledge Push, and a Governance Process.

The Economic Bargain includes agreements on international mechanisms for the equitable distribution of wealth. Transfers are tied to voluntary reductions in family size in countries with fast-growing populations, and to meeting globally agreed environmental targets. Technology transfer and joint sustainable development initiatives usher in a new era of cooperation between rich and poor regions. The Knowledge Push includes the formation of a formal global research network, which establishes centers around the world and works in close coordination with national and business research systems. The Governance Process explores fresh approaches for harmonizing economies with sustainability goals, and global governance with local prerogatives and responsibilities.

High-income regions speed the process of dematerialization in several ways. Services increase as a share of economic activity, a technology transition radically reduces resource intensity and consumerism wanes. In developing regions, material throughput increases initially in the push to satisfying basic needs, but stabilizes in the rapid convergence toward the moderate standards of the wealthier countries. Nowhere are the high consumption levels of late twentieth-century industrial societies again reached. Material sufficiency becomes the preferred lifestyle, while ostentatious consumption is viewed as primitive, a sign of vulgarity, and bad taste.

Some transnational corporations accept – and even advocate in some notable cases – the regulations and codes of conduct. Antisocial corporate behavior is discouraged by thorough public disclosure of information and a highly vigilant public. The polluter-pays principle is universally implemented. Firms that embrace the new business ethic of eco-efficiency and social responsibility gain competitive advantages. Clean technologies flourish as public preferences and prices shift. Various policy mechanisms promote the sustainability program, including a fiscal system based on eco-taxes that discourage environmental "bads." Well-designed environmental, economic, and social indicators measure the effectiveness of policies, providing a transparent basis for individuals and civil organization to monitor performance and seek change.

A process of social and economic renewal unleashes a spiral of positive change in developing countries. The quality of life improves at rates unprecedented in the historical record. The reinvention of development rests on effective governance, vastly improved educational opportunity, and socially inclusive participation. Poorer regions pioneer technologies and development approaches that conform to their unique climate, resources, and religious and cultural traditions. Innovations flow increasingly from poor to rich regions.

Declining costs of information technology facilitate universal access to global communication, even in remote areas. Mobile phones are ubiquitous while advances in artificial intelligence lead to automatic language translation. Information technology accelerates development, expands cultural options, and promotes education. It also helps people cope with their environment; for example, as weather forecasting better anticipates floods and droughts.

Support of science and technology and fertilization across cultures yields a wave of innovation. New technology is oriented towards sustainability and product durability, as the short-term profit motive is supplanted by longer-term economic rationality. An intensive effort makes subsistence farming more sustainable by combining advanced and traditional techniques. An eco-efficient industrial system captures and reuses waste in a cyclic process. New techniques for managing ecosystems and watersheds increase their resilience. With strong attention to safeguarding environmental and human health, engineered biosystems produce critical resources, food, and medicine.

A new metropolitan vision inspires the redesign of urban neighborhoods. Integrated settlement patterns place home, work, shops, and leisure activity in closer proximity.

Automobile dependence is reduced radically, and a sense of community and connectedness is reestablished. The condition for this urban renaissance is the elimination of the urban underclass, the ubiquitous signal of social distress of the previous era. For many people, the town-within-the-city provides the ideal balance of a human scale with cosmopolitan cultural intensity.

Dispersed small towns also become popular as communication and information technology allows for the decentralization of activities. The migration from rural to urban areas reverses as many opt for the lower stress level and increased contact with nature offered by smaller communities. A new spirit of community identification and participation is reinforced by changes in physical design, including decentralized renewable energy systems, greater reliance on local goods and restoration of environments. With attractive urban and rural alternatives, the mall culture fades as new urban and rural options underscore the sterility, hidden costs and isolation of suburbia.

Global governance is based on a federation of regions. In this nested structure, regions, nations, and communities have considerable control over socioeconomic decisions and approaches to environmental preservation, constrained only by the impacts on larger-scale environmental, economic, and social processes. Local energy systems vary, but must meet per-capita greenhouse gas emissions guidelines that are set by global agreement. Water strategies are tailored to local need, but must be compatible with allocation rules and ecosystem goals set at the river basin level.

A rejuvenated United Nations effectively fosters cooperation, security, and environmental health. Disputes and conflict are addressed by negotiation, collaboration, and consensus, with formal legal determinations required only rarely. A global peacekeeping force is called to hot spots on occasion, but this grows less frequent as the transition takes hold. As armies are cut back and defense systems dismantled, a massive peace dividend is channeled toward sustainability and poverty reduction.

In the new economic arrangements, markets are used to achieve production and allocation efficiency, but within the limits of non-market constraints defined by social, cultural, and environmental goals. These goals are set and implemented throughout the nested governance system with the participation of a wide range of stakeholders drawn from business, government, and civil society. Measures of development success increasingly focus on the quality of life and the state of the environment, rather than the discredited metric of economic growth. The time-horizon for economic decisions is lengthened to decades to take meaningful account of ecological processes.

Economic development continues indefinitely, mostly concentrated in the non-material realms of services, culture, and research. A labor-intensive craft economy rises spontaneously on the platform of the high technology base, providing a rewarding outlet for creative expression and a dizzying diversity of highly esthetic and treasured goods. Population growth slows and then stabilizes at relatively low levels as poverty is eliminated and women become equal participants in the life of communities. Universal access to education is critical to the voluntary reduction in family size.

Inherited environmental problems are abating, though some effects linger for many decades. Stabilized populations, less materialistic lifestyles, and a renewable energy transition sharply reduce greenhouse gas emissions and the climate begins to stabilize. Chemical pollution is virtually eliminated with the gradual phase-in of clean production processes. Sustainable agriculture practices are adopted universally, tailored to local conditions and tightly linked to local markets. This contributes to the moderation of water demands, decrease of freshwater stress, and restoration of ecosystems.

The exhilaration of pioneering a socially and environmentally superior way of life becomes a powerful attracting force in its own right, a self-fulfilling prophecy able to draw the present forward. An engaged citizenry rises to the challenge of forging a sustainable civilization of unprecedented creativity, freedom, and sense of shared destiny. But, of course, history continues to vex and surprise. The dreams of early idealists are defied to some extent by the persistence of pockets of poverty, the flaring of occasional geopolitical conflict and a global planet that is not yet totally healed. But people look back with justifiable pride on the immense achievements of their grandparents and great grandparents. The historic possibilities of their moment were seized. They bequeathed an era rich in human development, global solidarity, and ecological resilience.

Chapter 8

Reflections at the branch point

In the coming decades, the character of planetary development will be defined. Somehow, the spectacular growth of the human enterprise over the past two centuries will be reconciled with the Earth's finite resources and fragile environments. Somehow, the complex social and economic tensions posed by advancing global interdependence will be resolved. The question is, how? The shape of the global future will depend on how conditions unfold and humanity responds. The spectrum of possibilities is broad – world development could veer toward ecological and societal impoverishment or transition to a sustainable, just, and civilized future.

Planetary crossroads

The sweeping historical transformations of the modern era have brought us to the cusp of the planetary phase. The global system unfolds before us in a complex and turbulent process that could branch in many directions. The spiral of change in technology, values, knowledge, and institutions launched by the industrial revolution continues in new forms. The expansion toward a world system inevitably generated new and vexing social, economic, and environmental challenges. The transition has begun, but is yet to be completed.

Celebrants of global capitalism anticipate an endless cornucopia of goods, investment, and innovation. A permanent revolution in goods, consumer appetites, and human ingenuity would renew prosperity. The rising tide of wealth would raise all boats. The market would induce technical solutions for environmental problems. Critics find such celebration premature in light of the perilous global trends that weaken social cohesion and ecological sustainability. But even if the dangers of market-driven development were mitigated, the concern of many would remain: the global emporium of the future may not be worth living in.

The modern notion of sustainability crystallizes the sense that the drift of human history requires fundamental realignment. Humanity has the power to irreversibly alter the global ecology and reduce the planet's biological patrimony. It can no longer ignore the risks that unbridled growth and myopic policies place on the future. Sustainability recognizes that the connected global system couples the fate of humanity to that of nature and the fate of the privileged to that of the poor.

The capacity to transform the conditions of the future brings a new ethical imperative – to make choices today that are compatible with the well-being of future generations and the wider community of life. Once a pastime of visionaries and dreamers, taking the long view of the future has become essential for those concerned about the requirements today for a transition to sustainability tomorrow. Political leaders, research scientists, and citizens everywhere are challenged by the new imperative to think globally, holistically, and futuristically.

Global change is accelerating. The risk mounts that world development could drift into a condition of diminished resilience to shocks and surprises. The danger of slipping into irreversible trajectories jeopardizes the prospects for a transition to sustainability. When new technologies and ways of living "lock in," course changes become more difficult. To delay the start of the sustainability transition is to lose opportunities, with potentially fateful consequences. The precise timing and character of the danger points in this complex process are neither knowable nor predictable. What is clear is that it is not too late now to reduce environmental threats and social tensions by redirecting market-driven development with humanistic and ecological principles.

The path of reform

Poverty and inequity are not inevitable features of the modern world. The institutions, power relations, and assumptions that govern development reproduce them daily. Global wealth and global misery have both reached their pinnacles in the current era. This contradiction is the outcome of specific choices and circumstances that are subject to revision. The challenge is not so much to increase output, but to finally solve the problem of scarcity through fair distribution of resources, know-how and entitlements.

Great poverty amidst aggregate prosperity is one contradiction of the current era. The incompatibility of the growing scale of human activity with the finite capacity of the planet is a second. Was a global environmental crisis inevitable? Certainly the pursuit of immediate gain is a deep human propensity that was forged through eons of biological and cultural evolution and etched in values, norms, and institutions.

Technological prowess, population expansion, and the emergence of the capitalist growth machine were powerful engines in the march toward planetary limits. But these are preconditions and antecedents. The contemporary environmental problem is also the result of a set of historically contingent choices for production and consumption patterns.

Poverty, inequity, and ecological degradation are deeply embedded in contemporary institutional mechanisms and mindsets. They are stubborn, but not immutable. The *Policy Reform* scenario provides a concrete illustration of how the curve of development could be bent toward sustainability goals. It shows how government-led initiatives could set and meet social and environmental goals. While the goals are daunting, the necessary technologies, global wealth, and policy instruments are available for reducing the ecological stress and human poverty. Even without assuming alterations of conventional consumerist values and lifestyles and the evolution of global capitalism, gradual changes in social, technological, and resource-use patterns can become cumulatively significant.

Policy Reform represents the tacit vision of much of the sustainable development discussion expressed through global conventions, international conferences, and key strategic reports (WCED 1987). It would wed the goals of poverty reduction and environmental preservation with conventional assumptions on economic growth, corporate globalization, and consumerist values. The tensions inherent in such a union define the central challenge to the reform path to sustainability. In this vision, lifestyles, governance systems, and economic structures evolve slowly, but retain their fundamental character. Change comes as a gradual process of adjustment that constrains and steers market-driven development toward environmental and social targets. This program would require no less than the ascendancy of a new politics of global responsibility, long-range foresight, and sustainability.

Naturally, such a significant change in political philosophy and governance will be resisted. The long timeframe of sustainability conflicts with the short timeframe of special interest. The degree to which individuals, firms, and nations define their interests narrowly is a measure of the challenge. Myopia, parochialism, and selfishness are at odds with the long view, global sensitivity, and social orientation that must underpin a politics of sustainability. The implementation of comprehensive reforms requires a politics on behalf of the well-being of the world's poor, future generations, and global ecosystems. But conventional development stimulates individual accumulation, bottom-line thinking, and consumerism.

The dialectic of political change – public pressure from below and leadership from above – would need to overcome resistance by building a new vision and consensus for a politics of sustainability. If environmental and social tensions become aggravated in the years to come, the media, educators, and a widening circle of concerned businesses could foster greater awareness. Civil society acting through cultural and spiritual groups and NGOs could turn up the political pressure for sustainability reforms. In such a context, one might imagine the political leadership emerging for building the institutions, policies, and technologies for setting and pursuing sustainability goals could find political support.

The *Policy Reform* scenario is defined by environmental and social goals and a set of technological and policy strategies for meeting those goals. A host of policy instruments are available for steering development in the desired direction. The array of options for attaining social objectives is summarized in Table 8.1, and for environmental objectives in Table 8.2.

Table 8.1 Policy instruments to attain the proposed social objectives

- *International income transfers* through multilateral or bilateral official development assistance, special international funds, and voluntary donations mobilized by UN agencies, charities, NGOs, and others.
- *International agreements to safeguard the rights of the poor* from, for example, child exploitation and trade, labor and migration discrimination.
- *National income transfers* through public taxes and subsidies, and voluntary donations mobilized by charities, NGOs, and others.
- *Economic growth stimulation* in less developed countries through international assistance programs, trade arrangements, capacity building, and a variety of national structural adjustment efforts (note that social objectives are addressed only to the extent that increased income inequality does not negate the benefits of overall growth accruing to the poor).
- *Social empowerment measures* to recognize the rights of women, minorities, and children.
- *Investment in people* ("human capital") through education, health care, and training.
- *Participation by new social actors* through encouragement of communities, grass roots organizations, and NGO initiatives, and provision of mechanisms for taking part in decision-making.
- *Access of the poor to productive assets and income-generating opportunities* through a variety of steps, including land and natural resources tenure, technology transfer, infrastructure investment such as transport and storage facilities, and credit and marketing services targeted for poor producers.
- *Solidarity policies* ("safety nets") for disadvantaged groups such as destitute households, poor children, the sick, the handicapped, and the elderly.
- *Persuasion efforts* to promote solidarity values conducted by governments, religious groups, NGOs, advocacy organizations, and others through formal and informal education, information provision, public debate, media programming, etc.

Table 8.2 Policy instruments to attain the proposed environmental objectives

- *International agreements* on treatment of global commons, biodiversity, and resource and technology transfers.
- *Environmental codes of conduct* associated with international commerce, investment, tourism, transport, migration, biodiversity uses, etc.
- *Reactive regulations, laws and rules* that discourage known environmental disruption (e.g., regulation to control pollution and natural resource depletion).
- *Proactive regulations, laws and rules* that encourage the introduction of sustainable practices (e.g., measures that foster energy efficiency, recycling and clean technologies, biodiversity, and natural resource conservation).
- *Scientific and technological development* for (a) pollution prevention, (b) clean production and consumption technologies, (c) sustainable use of natural environments and resources, (d) deepening the understanding of environment and society interactions, and (e) enhancing decision-making tools (e.g., green accounting, long-range assessment, EIA).
- *Green budgets* which use fiscal measures as a lever to promote environmental goals, including: (a) directing public expenditure to ecologically-friendly products (e.g., efficient technologies, recycled products), (b) using pollution and natural resource depletion taxes ("eco-taxes") as a primary source of public revenue, and c) moving subsidies from activities that foster resource inefficiency and environmental degradation to those that promote sustainability.
- *Social empowerment policies* that acknowledge and support traditional property rights of local and indigenous communities and traditional sustainable uses of biodiversity.
- *Persuasion efforts* to promote environmental sustainability values conducted by governments, religious groups, NGOs, advocacy organizations, and others through formal and informal education, information provision, public debate, media programming, etc.
- *Economic policies* to motivate sustainable behavior based on individual and corporate economic self-interest, and to use market allocation mechanisms to attain environmental goals in an economically efficient manner, including:

  - *Full pricing of environmental goods and services* ("internalization of externalities") through (a) charges or royalty fees on pollution, production, resource extraction, or resource development, (b) ecological taxes, (c) deposit-refund systems, bond and deposit systems, and (d) import tariffs and export taxes.
  - *Stimulation of environmental markets* through (a) tradable permits for pollution and resource use, (b) assigning property rights to own, use, or develop resources, and (c) liability and insurance markets.
  - *Distribution of subsidies* through (a) tax breaks, (b) tax differentials, (c) exemptions from charges, (d) grants, (e) soft loans, (f) accelerated depreciation allowances, etc.

Each of these broad policy areas includes an enormous variety of formulations. It is encouraging that many of these instruments currently are being employed at local, national, or international levels. Yet, existing policy lacks the breadth and depth required for a transition to sustainable development. While success stories exist – the international ban on ozone-depleting substances, the worldwide reduction in lead use, the significant increase in life expectancy among developing countries' populations – the process of change at the scale of a *Policy Reform* scenario has hardly begun.

There is no single best package of options. Rather, diverse approaches that are appropriate to the issue, level of governance, and local concerns will be needed. A number of considerations come into play in effective policy design. These include political acceptability, economic efficiency, equity impacts, and cultural preferences. Yet, across the diversity of policy approaches, we can identify some important guidelines for formulating integrated sustainability policies. Foremost is the need for integration and consistency across sectors, scales, and time frames.

A complex and interactive set of sectoral processes drives socio-ecological systems. Environmental, economic, technological, social, cultural, and political dimensions are mutually conditioned on one another. If they are fragmented and contradictory, sectoral policies can negate one another – initiatives to reduce poverty might exacerbate environmental degradation, and vice versa. By contrast, an integrated framework seeks reinforcement and synergy between sectoral policies. Measures to protect ecosystems and provide sustainable livelihoods can reinforce water and energy sectors. So can initiatives in agriculture.

Policies need to be consistent across spatial scales, as well as across sectors. Different sustainability issues come into focus at global, regional, national, and local levels. An integrated perspective is essential for formulating goals and policies at each level, and for identifying synergies and incompatibilities. A nested policy structure for sustainability would reflect the multilevel character of environmental, social, and economic problems.

An additional dimension of integration is time – the harmonization of short- and long-term outlooks. A long horizon is an obvious requirement for policies that aim to affect development over many decades. But long-term policy is often, by default, the accumulation of a series of actions based on more immediate considerations. Since alternative policies can often address short-term goals, but have very different long-term impacts, they must be designed with the future in mind. The scenario approach is particularly useful in helping incorporate long-term considerations into today's policy discussions.

Flexibility and adaptability to change are also critical attributes of policy formulation because of the many uncertainties inherent as socio-ecological systems unfold. Rigid and bureaucratic processes are not well matched for reflecting these uncertainties and adapting to new information. While governments shoulder the responsibility for policy reform, participatory processes can mobilize relevant institutions and stakeholders, build consensus and benefit from their diverse perspectives and knowledge. They also provide a basis for monitoring and seeking revision in the course of time. Scenario exercises can reveal contrasting visions and interests and promote a pluralistic dialogue.

Science has much to contribute to policy reform. Sound analysis that is effectively communicated can support public processes, decision-making, and the formulation of policy. In recent decades, society has increasingly turned to science for insight into environmental dilemmas, and the scientific professions have responded. Conventional disciplinary boundaries have been transcended to accommodate the interdisciplinary requirements of complex ecological problems. The interface between science and society is evolving. Research priorities for a sustainability transition are under active discussion within the scientific community (BSD 1998, Kates *et al.* 2001). The challenge is to advance understanding of the dynamics of socio-ecological systems. New approaches must recognize the inherent uncertainty and normative character of sustainable development problems. Urgent action is needed to develop methods, train a new cadre of sustainability professionals and build institutional capacity worldwide. The task is no less than forging a new science of sustainability.

Turning toward a new paradigm

The incrementalism of *Policy Reform* seeks sustainability without revising the underlying development paradigm of consumerism and the centrality of economic growth. But would it ultimately fulfill human aspirations? The *Great Transition* vision transcends the conventional paradigm to link sustainability with a new development vision. The search for

Table 8.3 Sustainability pushes and pulls

Pushes	Pulls
Insecurity as socio-ecological crisis matures	Economic security and development
Concern that policy adjustments are insufficient	Sense of meaning and purpose
Fear of authoritarianism and barbarization	Better lifestyles and greater fulfillment
Alienation from dominant culture	Community and political engagement
Stressful lifestyles	Connection to nature
Anxiety about the future	Contributing to a sustainable future

environmental and social resilience finds expression in alternative values and the search for a new basis for human fulfillment.

The motivations for a *Great Transition* fall into two categories – the "push" of necessity and the "pull" of desirability (Table 8.3). The push is the fear that current trends are unsustainable and that they augur future crises. This is the motivation for *Policy Reform*, as well. The pull is the lure of more desirable lifestyles and lifeways.

Great Transitions would revise the concept of wealth. On a base of adequate material resources, the stress turns to cultural, community, and spiritual richness. Most of human history was dominated by the struggle for survival. Primitive economies evolved to provide sustenance and a modest surplus, and to cope with harsh and unpredictable natural forces. The long cultural journey through tool making, agriculture, and modern technology traversed a path from human scarcity and vulnerability to abundance and domination of nature.

The immense growth in aggregate economic output has finally put the possibility of a post-scarcity world on the historical agenda. The precondition for a *Great Transition* is the capacity to provide a comfortable and sufficient standard of living for all. On this basis, the vision of a better life can legitimately and inclusively emphasize meaning, purpose and happiness. These can be largely satisfied through nonmaterial means such as creative arts, research, human relationships in their many forms, and appreciation of nature.

Alternative scenarios of the future are based on different outcomes of a set of causative drivers that have psychological, cognitive, and institutional dimensions. To illuminate the *Great Transition* further, it is useful to decompose these into *proximate* and *ultimate* drivers. Proximate drivers are directly responsible for propelling the scenario. Ultimate drivers condition the proximate drivers.

For example, water pollution from factories depends proximately on the process technology used, the volume and pattern of industrial activity, and prevailing regulation and economic incentives. Ultimately, these depend on the value society puts on environmental quality, the lobbying power of interested parties, consumption patterns that set the demand for the industry's products, and the scientific understanding of the causes, impacts, and dynamics of water pollution. To take a second example, hunger in the developing world is driven proximately by a lack of access to food, meager incomes, and disempowerment. These are ultimately shaped by values, class, and power.

A framework is introduced in Figure 8.1. The "critical trends" are associated with the *Market Forces* scenario. The "proximate drivers" draw attention to the direct levers of change that define *Policy Reform* variations. The "ultimate drivers" refer to the shape of the fundamental structure of values, knowledge, and empowerment that, if suitably altered, could initiate a *Great Transition* scenario. Because proximate drivers have a direct influence on trends, they are subject to short-term policy intervention. The more stable ultimate

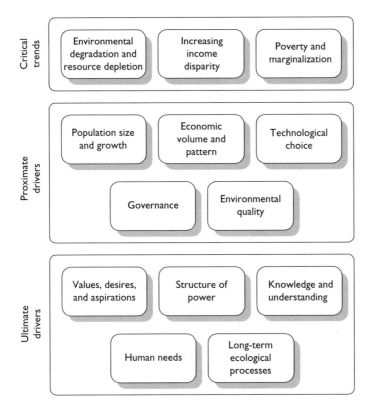

Critical trends

Environmental degradation and resource depletion

Increasing income disparity

Poverty and marginalization

Proximate drivers

Population size and growth

Economic volume and pattern

Technological choice

Governance

Environmental quality

Ultimate drivers

Values, desires, and aspirations

Structure of power

Knowledge and understanding

Human needs

Long-term ecological processes

Figure 8.1 Trends and drivers.

drivers are subject to gradual cultural and political processes. But they also are influenced by acute events, such as environmental or geopolitical crises.

The major proximate drivers include:

- *Population size and growth.* Population size is a major determinant of total pressure on the environment and resources. Population density often correlates to local environmental pressure. Rapid rates of population growth tax the capacity of governments to provide adequate infrastructure and services. The stresses are manifest in developing regions in both burgeoning urban centers and zones of rural poverty.
- *Economic volume and pattern.* All else equal, the larger the economy, the greater the throughput of material resources and energy. Levels of natural resource consumption and waste production tend to correlate with the level of economic output. The detailed relationship is closely linked to the pattern of consumption, lifestyles, income distribution, and poverty.
- *Technological choice.* The choice of technology – the means by which societies, organizations, and individuals accomplish tasks – influences the environmental and social impact per unit of production or consumption. Eco-efficient production and consumption

technologies play a major role in formulating *Policy Reform* scenarios. The choice of technology depends on engineering criteria, cost, and cultural preference.

- *Governance.* Construed broadly, governance is the aggregate of processes and arrangements by which human communities interact. It includes both governmental and nongovernmental mechanisms. Institutional and legal fragmentation is a major obstacle to the integrated policy response required for sustainable development. Allocation mechanisms contribute both to environmental degradation and to poverty. Global systems of governance are struggling to cope with the immense challenges of economic globalization and transnational environmental threats.
- *Environmental quality.* Poor soils, extreme climate, and fragile ecosystems may be due to natural conditions and may be exacerbated by anthropogenic environmental degradation. Many of the rural poor live in resource-scarce areas, while the urban poor tend to live in the most polluted areas. Impoverishment and low environmental quality can combine in a downwardly spiraling process. Human action can address certain aspects of environmental quality through environmental restoration efforts and the reduction of pressure on environmental systems.

The ultimate drivers in Figure 8.1 are:

- *Values, desires, and aspirations.* The prevailing values in a society set the criteria for individual ambition, ethics, and aesthetic sensibility. Values are culturally conditioned, defining the social consensus on what is considered normal or desirable. A society will lie somewhere along a spectrum between a sense of common purpose and antagonism, between social solidarity and individualism, between materialism and a concern for deeper meaning. Consumerism is significant in driving human aspirations in directions that appear to be unsustainable. A sense of solidarity with both current and future generations, were it to become a defining global value, would form the motivational basis for sustainable development.
- *Structure of power.* The control of resources, influence, and decision-making defines the power structure. The structure circumscribes the limits of the possible, particularly for the weak, and sets priorities for society. It defines allocation and redistribution patterns. Through media control it can shape human values, preferences, and consumption patterns. The current global power structure is notable for its pronounced international asymmetry.
- *Knowledge and understanding.* The scientific enterprise – the formal organization of knowledge into theory and observation – shapes the worldviews of modern societies and the capacity to manipulate natural and social systems. Scientific advance in understanding the complexity, holism, and normative character of the sustainability problem could be a driver of positive change. In addition, informal knowledge – the manner in which people perceive and make sense of their conditions as individuals and participants in society – influences behavior and responses to changing circumstances. Culture, education, and the media shape understanding.
- *Human needs.* These include material needs, such as for food and shelter, as well as nonmaterial needs, such as love and meaning. Fundamental human needs – the generic human requirements for health – are universal. But the specific elements for satisfying them are to a significant degree culturally determined. Policies can affect the choice of specific "satisfiers." Furthermore, advertising and the media can influence the creation of wants and their conversion into felt needs.

- *Long-term ecological processes*. Ecological processes mold the physical setting against which human trends and scenarios play out. These include long-term evolutionary processes, short-term disturbances such as hurricanes and volcanic eruptions and natural cycles such as El Niño events. The human footprint on the environment is significantly disturbing these global ecological processes. Notable are the severe biodiversity loss and the perilous modification of the global climate system.

Policy Reform scenarios examine the instrumental changes that are needed without altering fundamental values and institutions. The focus is on redirecting the proximate drivers through economic reform, regulation, redistribution, and technological initiative. Whether this would suffice is an open question. The ultimate drivers concern the basic character of human motivation and social structure. The greater challenge of addressing ultimate drivers is left to *Great Transitions* scenarios. Efforts to move toward such a future would need to address dominant values, the character of knowledge and the prevailing structure of power.

A world to choose

The long-range future cannot be predicted mechanically from current conditions and driving forces. The arc of history can branch in fundamental ways. Scenarios have allowed us to gauge the prospects and risks of a spectrum of global development paths. *Market Forces* holds the potential to generate great wealth, but risks perpetuating poverty, environmental degradation, and conflict. *Policy Reform* shows that governmental initiatives could address these issues if the political will emerges. *Barbarization* envisions the catastrophic hazards if it does not. *Great Transition* depicts how the search for sustainability and fulfillment might lead to an alternative global path as dominant values change.

It is natural that policy-oriented scenario analyses focus, in the first instance, on *Conventional Worlds*. This is where the official discussion occurs about sustainable development within governmental processes, Earth Summits, and international negotiations. The issue is reform, not transition. But *Conventional Worlds* scenarios cannot be assumed to be likely. Nor can historic tilts toward *Barbarization* or *Great Transition* worlds be ruled out. Indeed, one need not be a cynic to observe troubling portents of the former in the evolving social fissures and ecological uncertainties of our time. Nobody can rigorously assign probabilities to such transformations since they depend on inherent uncertainties in complex systems – and on human choices that are yet to be made.

In *Market Forces*, the dominant driver of change is the expansion of unfettered global markets. This is perhaps the dominant tendency today as new technologies and institutions advance the borderless economy and the cultural changes it entrains. Yet, we find that this future could be undermined by its own socio-ecological perils and contradictions. The lesson is not that the world will inevitably follow an unsatisfactory development path in the absence of proactive policy interventions. Even without action, it is possible that unforeseen surprises, feedbacks and adaptations could mitigate the social and environmental pressures of the *Market Forces* world – or exacerbate them. The message is, rather, that there are deepening risks and uncertainties, and that complacency is not a valid option for those concerned about passing on a more secure future.

The discussion of *Policy Reform* scenarios brought both good news and bad news. The good news is that great strides toward sustainability can be achieved without either a social revolution or the *deus ex machina* of a technological miracle. Though there are numerous constraints and challenges, the cumulative effects of a comprehensive family of feasible incremental technological and social policy adjustments can make a substantial difference. An evolutionary process that promotes appropriate technology, environmental management, and greater international and social equity would take us a long way.

The bad news is that *Policy Reform* does require a strong postulate – the hypothesis of sufficient political will. It assumes that an unprecedented governmental commitment arises for achieving sustainability goals, and that effective economic, social, and institutional reforms are introduced for achieving them, in the context of the persistence of conventional values and institutions. The hypothesis poses critical questions for the *Policy Reform* route to the sustainability transition. Can the political basis for constraining and guiding market development emerge in a world where consumerism and individualism are dominant values? Can the resistance of special interests, the myopia of narrow outlooks, and the inertia of complacency be overcome? Does a sustainability transition require a deeper reconsideration of the conventional development paradigm?

Barbarization is a somber reminder of the costs of the failure of development. Understanding the possible pathways to *Barbarization* can have fundamental implications for policy, as well. Though not yet in the realm of normal policy discourse, greater appreciation of the risk of social polarization and environmental crises can stimulate actions to mute critical tensions in the system. It would be highly imprudent to assume that the internal socioeconomic and biophysical tensions of a conventional development world do not pose serious risks to the health of socio-ecological systems. Awareness of the pathways to rapid destabilization is a prod to action. The problem of critical uncertainties and global discontinuities invites a major program of research, education, and policy formulation.

The burning issue is whether this awareness and a mobilization for sustainable development can, at a minimum, reduce environmental disruption and social destabilization. Positive incremental adjustments in global development can keep open the opportunity for deeper processes of social renewal by buying time and spreading the paradigm of sustainability and global responsibility. Then perhaps the value changes driving a *Great Transition* can begin to coalesce.

That would seem to require that voices representing a wide range of insights, aspirations, and interests gradually join in a coherent global movement. The twenty-first century will be characterized by an evolving planetary system that binds nations and people ever more tightly in a shared ecological and political destiny. These conditions could find their counterpart in a planetary zeitgeist of sustainability and solidarity, the path leading to a new global development paradigm. The weighty challenge for current generations is to think and act in ways that reduce social and ecological stress, while keeping future opportunities open for a *Great Transition*.

The story of the future has yet to be written. Rather than simply "happening," the human future depends on collective and individual choices and actions. At the same time, physical and societal conditions circumscribe the range of possible futures. The future path actually taken will depend on the interplay between historical constraints and human responses – between necessity and freedom.

Hegemonic values, dominant institutions, and habits of behavior have great inertia that resists historical forces for change. Under relatively stable conditions, human beings have displayed little appetite for significant alteration of behavior in anticipation of possible but uncertain crises in the future. Rather, the tendency is to maintain old patterns until a crisis forces reaction. This would seem to consign to futility any appeal for a voluntary redirection of global development.

However, in periods of instability, where dominant social processes become less binding, "choices" can ripple through the system in a kind of social tsunami. At such historical branch points, institutional constraints abate, conventional wisdom is questioned, and personal priorities are reconsidered. Collective and individual decisions have greater potential to affect events. Mounting environmental, social, and geopolitical problems augur distant risks and cause apprehension about the future. Growing global turbulence is the context for the spread of foresight, popular awareness, and political change.

Voices of concern are everywhere, often finding expression through the thousands of local organizations that take up a dizzying range of specific environmental and social issues. The capacity of these groups to educate, articulate demands, and foster change is growing as the communication revolution provides important new media for social mobilization. Countless others, concerned about the future, are isolated and lack the context for sustained reflection and action on the dangers and possibilities posed by the advancing world system.

A compelling framework is needed as a basis of global unity. It must be rich in history, science, and vision. It must illuminate the complex factors shaping global development. It must reveal the unity that underlies diverse concerns and values. In the absence of such a shared perspective, the fragmentation of knowledge, understanding, and political expression will continue to enfeeble and isolate the forces for a progressive change of direction. Advocates for specific problems, group interests and local issues are necessary actors in the global transition. But a sufficient movement for this task will require a holistic vision of its historic challenge and possibility.

The "celebration of diversity" has become a liberating theme of the last 30 years. The "politics of identity" and "thinking globally, acting locally" supersede the stultifying top-down ideologies of the past. Now there is a need for unity. Diverse and dispersed individuals and organizations must link their activities, worldviews, and goals to guide complementary and coordinated action. In the absence of such unification, it is difficult to see how the politics of sustainability can transcend its current political liabilities – lip service at the top and fragmentation on the ground.

The destinies of rich and poor, entrenched and excluded, North and South, current and future generations, are becoming ever more tightly linked – there can be no separate solutions in the search for ecological and social renewal. If discussion and organization rise to the level of the human species, the biosphere and global society, a sustainable and desirable planetary transition can be envisioned.

Annex

D-1 Population

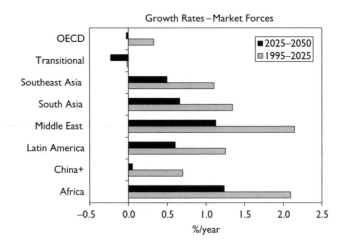

Growth Rates – Market Forces

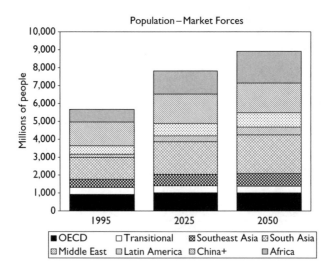

Population – Market Forces

D-1 Population

	Population (millions)			Growth rate (%/year)			Index (1995 = 1)	
	1995	2025	2050	1995–2025	2025–2050	1995–2050	2025	2050
Market Forces								
Africa	697	1,298	1,766	2.1	1.2	1.7	1.9	2.5
China+	1,330	1,639	1,660	0.7	0.1	0.4	1.2	1.2
Latin America	476	692	804	1.3	0.6	1.0	1.5	1.7
Middle East	173	327	433	2.1	1.1	1.7	1.9	2.5
South Asia	1,228	1,834	2,162	1.3	0.7	1.0	1.5	1.8
Southeast Asia	453	630	713	1.1	0.5	0.8	1.4	1.6
Developing	**4,357**	**6,420**	**7,538**	**1.3**	**0.6**	**1.0**	**1.5**	**1.7**
Eastern Europe	99	94	83	−0.2	−0.5	−0.3	0.9	0.8
FSU	292	295	284	0.0	−0.2	−0.1	1.0	1.0
Transitional	**391**	**389**	**367**	**0.0**	**−0.2**	**−0.1**	**1.0**	**0.9**
North America	301	368	397	0.7	0.3	0.5	1.2	1.3
Pac OECD	150	153	142	0.1	−0.3	−0.1	1.0	0.9
Western Europe	468	492	467	0.2	−0.2	0.0	1.1	1.0
OECD	**919**	**1,013**	**1,006**	**0.3**	**0.0**	**0.2**	**1.1**	**1.1**
World	**5,667**	**7,822**	**8,911**	**1.1**	**0.5**	**0.8**	**1.4**	**1.6**
Policy Reform								
Africa	697	1,272	1,678	2.0	1.1	1.6	1.8	2.4
China+	1,330	1,607	1,577	0.6	−0.1	0.3	1.2	1.2
Latin America	476	678	764	1.2	0.5	0.9	1.4	1.6
Middle East	173	321	411	2.1	1.0	1.6	1.9	2.4
South Asia	1,228	1,797	2,054	1.3	0.5	0.9	1.5	1.7
Southeast Asia	453	618	677	1.0	0.4	0.7	1.4	1.5
Developing	**4,357**	**6,293**	**7,161**	**1.2**	**0.5**	**0.9**	**1.4**	**1.6**
Eastern Europe	99	92	79	−0.2	−0.6	−0.4	0.9	0.8
FSU	292	289	270	0.0	−0.3	−0.1	1.0	0.9
Transitional	**391**	**381**	**349**	**−0.1**	**−0.4**	**−0.2**	**1.0**	**0.9**
North America	301	368	397	0.7	0.3	0.5	1.2	1.3
Pac OECD	150	153	142	0.1	−0.3	−0.1	1.0	0.9
Western Europe	468	492	467	0.2	−0.2	0.0	1.1	1.0
OECD	**919**	**1,013**	**1,006**	**0.3**	**0.0**	**0.2**	**1.1**	**1.1**
World	**5,667**	**7,687**	**8,516**	**1.0**	**0.4**	**0.7**	**1.4**	**1.5**

D-2 Urbanization

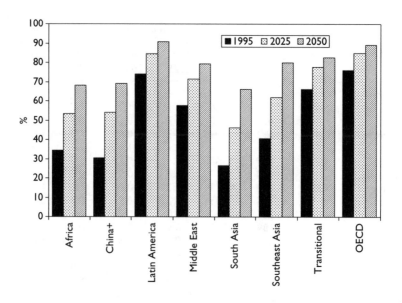

D-2 Urbanization

	Urbanization (%)			Urban population (millions)			Index (1995 = 1)	
	1995	2025	2050	1995	2025	2050	2025	2050
Market Forces								
Africa	34	53	68	239	694	1,206	2.9	5.0
China+	31	54	69	407	886	1,147	2.2	2.8
Latin America	74	85	91	353	586	730	1.7	2.1
Middle East	58	72	79	100	234	344	2.3	3.4
South Asia	27	46	66	326	846	1,431	2.6	4.4
Southeast Asia	41	62	80	184	390	571	2.1	3.1
Developing	**37**	**57**	**72**	**1,609**	**3,636**	**5,429**	**2.3**	**3.4**
Eastern Europe	62	76	80	61	71	66	1.2	1.1
FSU	68	78	83	198	231	237	1.2	1.2
Transitional	**66**	**78**	**83**	**259**	**302**	**303**	**1.2**	**1.2**
North America	76	85	88	229	312	348	1.4	1.5
Pac OECD	78	85	90	117	130	128	1.1	1.1
Western Europe	76	85	90	354	419	421	1.2	1.2
OECD	**76**	**85**	**89**	**700**	**861**	**897**	**1.2**	**1.3**
World	**45**	**61**	**74**	**2,568**	**4,799**	**6,629**	**1.9**	**2.6**
Policy Reform								
Africa	34	53	68	239	680	1,145	2.8	4.8
China+	31	54	69	407	869	1,090	2.1	2.7
Latin America	74	85	91	353	574	693	1.6	2.0
Middle East	58	71	79	100	229	326	2.3	3.3
South Asia	27	46	66	326	829	1,359	2.5	4.2
Southeast Asia	41	62	80	184	382	542	2.1	2.9
Developing	**37**	**57**	**72**	**1,609**	**3,563**	**5,155**	**2.2**	**3.2**
Eastern Europe	62	76	80	61	70	63	1.1	1.0
FSU	68	78	83	198	226	225	1.1	1.1
Transitional	**66**	**78**	**83**	**259**	**296**	**288**	**1.1**	**1.1**
North America	76	85	88	229	312	348	1.4	1.5
Pac OECD	78	85	90	117	130	128	1.1	1.1
Western Europe	76	85	90	354	419	421	1.2	1.2
OECD	**76**	**85**	**89**	**700**	**861**	**897**	**1.2**	**1.3**
World	**45**	**61**	**74**	**2,568**	**4,720**	**6,340**	**1.8**	**2.5**

E-1 GDP

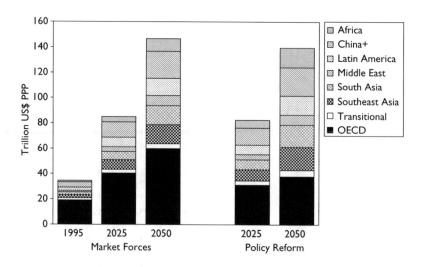

E-1 GDP

	GDP (trillion dollars)				Growth rate (%/year)			Index (1995 = 1)	
	MER	PPP							
	1995	1995	2025	2050	1995–2025	2025–2050	1995–2050	2025	2050
Market Forces									
Africa	0.5	1.4	4.4	9.7	3.9	3.3	3.6	3.2	7.1
China+	0.8	3.9	12.1	21.5	3.8	2.3	3.1	3.1	5.5
Latin America	1.7	2.8	7.3	13.6	3.3	2.5	2.9	2.6	4.9
Middle East	0.7	1.3	4.1	8.1	4.0	2.8	3.4	3.2	6.4
South Asia	0.4	1.9	6.6	14.8	4.3	3.3	3.8	3.5	7.9
Southeast Asia	1.2	2.5	7.5	14.9	3.7	2.8	3.3	3.0	5.9
Developing	**5.2**	**13.7**	**41.8**	**82.6**	**3.8**	**2.8**	**3.3**	**3.0**	**6.0**
Eastern Europe	0.3	0.6	0.9	1.2	1.8	0.9	1.4	1.7	2.1
FSU	0.6	1.1	2.0	2.7	2.0	1.2	1.6	1.8	2.5
Transitional	**0.9**	**1.6**	**2.9**	**3.9**	**1.9**	**1.1**	**1.6**	**1.8**	**2.3**
North America	7.6	8.0	18.2	28.4	2.8	1.8	2.3	2.3	3.6
Pac OECD	5.6	3.2	6.2	8.5	2.2	1.3	1.8	1.9	2.7
Western Europe	9.5	8.1	16.0	23.1	2.3	1.5	1.9	2.0	2.8
OECD	**22.7**	**19.3**	**40.3**	**60.0**	**2.5**	**1.6**	**2.1**	**2.1**	**3.1**
World	**28.7**	**34.7**	**85.1**	**146.5**	**3.0**	**2.2**	**2.7**	**2.5**	**4.2**
Policy Reform									
Africa	0.5	1.4	6.4	15.2	5.3	3.5	4.5	4.7	11.1
China+	0.8	3.9	13.1	22.8	4.1	2.2	3.2	3.3	5.8
Latin America	1.7	2.8	7.9	14.8	3.5	2.5	3.1	2.8	5.3
Middle East	0.7	1.3	4.1	8.2	4.0	2.8	3.4	3.3	6.4
South Asia	0.4	1.9	7.9	17.5	4.9	3.2	4.2	4.2	9.4
Southeast Asia	1.2	2.5	8.4	18.3	4.1	3.2	3.7	3.3	7.3
Developing	**5.2**	**13.7**	**47.9**	**96.7**	**4.2**	**2.9**	**3.6**	**3.5**	**7.0**
Eastern Europe	0.3	0.6	1.0	1.4	2.1	1.0	1.6	1.9	2.4
FSU	0.6	1.1	2.4	3.5	2.6	1.5	2.1	2.2	3.2
Transitional	**0.9**	**1.6**	**3.4**	**4.8**	**2.5**	**1.4**	**2.0**	**2.1**	**2.9**
North America	7.6	8.0	13.1	16.7	1.7	1.0	1.4	1.6	2.1
Pac OECD	5.6	3.2	4.8	5.4	1.3	0.5	0.9	1.5	1.7
Western Europe	9.5	8.1	13.5	16.0	1.7	0.7	1.2	1.7	2.0
OECD	**22.7**	**19.3**	**31.3**	**38.1**	**1.6**	**0.8**	**1.2**	**1.6**	**2.0**
World	**28.7**	**34.7**	**82.6**	**139.6**	**2.9**	**2.1**	**2.6**	**2.4**	**4.0**

E-2 Structure of GDP

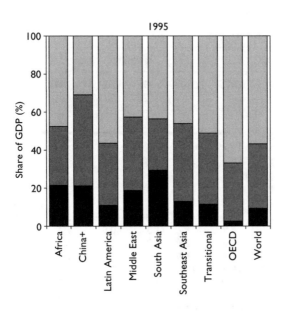

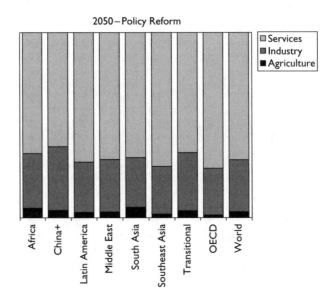

E-2 Structure of GDP

	Share of GDP (%)								
	1995			2025			2050		
	Agri-culture	Industry	Services	Agri-culture	Industry	Services	Agri-culture	Industry	Services
Market Forces									
Africa	21	31	48	13	31	56	8	30	62
China+	21	48	31	8	43	49	4	36	59
Latin America	11	33	56	6	31	63	3	28	68
Middle East	19	39	43	9	35	56	4	29	67
South Asia	29	27	44	13	27	60	7	27	66
Southeast Asia	13	41	46	5	36	59	3	26	72
Developing	**18**	**38**	**43**	**9**	**35**	**56**	**5**	**30**	**65**
Eastern Europe	10	38	52	6	36	59	4	32	64
FSU	12	37	51	7	36	57	5	34	61
Transitional	**12**	**37**	**51**	**7**	**36**	**58**	**5**	**33**	**62**
North America	2	26	72	1	24	75	1	21	78
Pac OECD	2	37	61	1	34	65	1	32	68
Western Europe	3	32	65	2	29	69	1	27	72
OECD	**3**	**31**	**67**	**1**	**28**	**71**	**1**	**25**	**74**
World	**9**	**34**	**57**	**5**	**32**	**63**	**3**	**28**	**69**
Policy Reform									
Africa	21	31	48	9	31	61	5	30	65
China+	21	48	31	7	42	50	4	34	62
Latin America	11	33	56	5	31	64	3	27	70
Middle East	19	39	43	8	35	57	3	28	69
South Asia	29	27	44	10	27	62	6	27	68
Southeast Asia	13	41	46	5	34	61	2	26	72
Developing	**18**	**38**	**43**	**7**	**34**	**58**	**4**	**29**	**67**
Eastern Europe	10	38	52	5	35	60	3	31	66
FSU	12	37	51	6	35	59	4	32	64
Transitional	**12**	**37**	**51**	**6**	**35**	**60**	**4**	**31**	**65**
North America	2	26	72	2	24	75	1	21	77
Pac OECD	2	37	61	2	34	65	1	32	67
Western Europe	3	32	65	2	29	69	2	27	71
OECD	**3**	**31**	**67**	**2**	**28**	**71**	**1**	**25**	**73**
World	**9**	**34**	**57**	**5**	**32**	**63**	**3**	**28**	**69**

E-3 Income

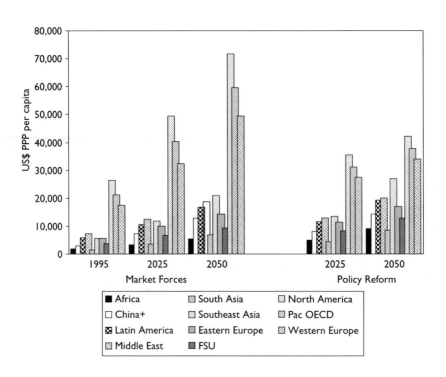

E-3 Income

	GDP per capita (1995 US$ PPP)			Growth rate (%/year)			Index (1995 = 1)	
	1995	2025	2050	1995–2025	2025–2050	1995–2050	2025	2050
Market Forces								
Africa	2,000	3,400	5,500	1.8	2.0	1.9	1.7	2.8
China+	3,000	7,400	13,000	3.1	2.3	2.7	2.5	4.4
Latin America	5,800	10,500	16,900	2.0	1.9	2.0	1.8	2.9
Middle East	7,300	12,400	18,700	1.8	1.6	1.7	1.7	2.6
South Asia	1,500	3,600	6,900	2.9	2.6	2.8	2.4	4.5
Southeast Asia	5,600	11,900	20,900	2.6	2.3	2.4	2.1	3.8
Developing	**3,200**	**6,500**	**11,000**	**2.4**	**2.1**	**2.3**	**2.1**	**3.5**
Eastern Europe	5,600	10,100	14,300	2.0	1.4	1.7	1.8	2.5
FSU	3,800	6,700	9,400	2.0	1.4	1.7	1.8	2.5
Transitional	**4,200**	**7,500**	**10,500**	**2.0**	**1.3**	**1.7**	**1.8**	**2.5**
North America	26,400	49,400	71,600	2.1	1.5	1.8	1.9	2.7
Pac OECD	21,300	40,300	59,700	2.2	1.6	1.9	1.9	2.8
Western Europe	17,400	32,500	49,400	2.1	1.7	1.9	1.9	2.8
OECD	**21,000**	**39,800**	**59,600**	**2.2**	**1.6**	**1.9**	**1.9**	**2.8**
World	**6,100**	**10,900**	**16,400**	**1.9**	**1.7**	**1.8**	**1.8**	**2.7**
Policy Reform								
Africa	2,000	5,000	9,100	3.2	2.4	2.8	2.6	4.6
China+	3,000	8,200	14,400	3.4	2.3	2.9	2.8	4.9
Latin America	5,800	11,600	19,300	2.3	2.1	2.2	2.0	3.3
Middle East	7,300	12,900	19,800	1.9	1.7	1.8	1.8	2.7
South Asia	1,500	4,400	8,500	3.6	2.7	3.2	2.9	5.6
Southeast Asia	5,600	13,600	27,100	3.0	2.8	2.9	2.4	4.9
Developing	**3,200**	**7,600**	**13,500**	**3.0**	**2.3**	**2.7**	**2.4**	**4.3**
Eastern Europe	5,600	11,400	17,100	2.4	1.6	2.1	2.0	3.1
FSU	3,800	8,200	12,900	2.7	1.8	2.3	2.2	3.4
Transitional	**4,200**	**9,000**	**13,800**	**2.6**	**1.7**	**2.2**	**2.1**	**3.3**
North America	26,400	35,500	42,100	1.0	0.7	0.8	1.3	1.6
Pac OECD	21,300	31,100	37,800	1.3	0.8	1.0	1.5	1.8
Western Europe	17,400	27,500	34,200	1.5	0.9	1.2	1.6	2.0
OECD	**21,000**	**30,900**	**37,800**	**1.3**	**0.8**	**1.1**	**1.5**	**1.8**
World	**6,100**	**10,700**	**16,400**	**1.9**	**1.7**	**1.8**	**1.8**	**2.7**

S-1 Income distribution

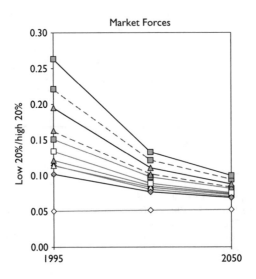

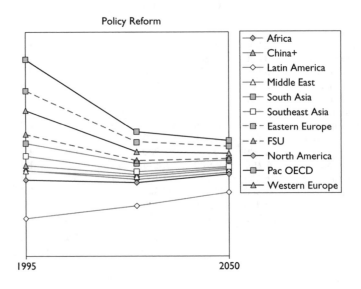

S-1 Income distribution

	National equity (low 20%/high 20%)			Index (1995 = 1)		Gini coefficient (%)		
	1995	2025	2050	2025	2050	1995	2025	2050
Market Forces								
Africa	0.11	0.08	0.07	0.7	0.6	43	48	49
China+	0.12	0.09	0.07	0.7	0.6	40	46	48
Latin America	0.05	0.05	0.05	1.0	1.0	55	54	54
Middle East	0.11	0.08	0.07	0.7	0.6	41	46	49
South Asia	0.15	0.10	0.08	0.6	0.5	37	44	47
Southeast Asia	0.13	0.09	0.08	0.7	0.6	39	46	48
Developing	**0.12**	**0.08**	**0.07**	**0.7**	**0.6**	**41**	**47**	**49**
Eastern Europe	0.22	0.12	0.09	0.5	0.4	30	40	44
FSU	0.16	0.10	0.08	0.6	0.5	35	43	46
Transitional	**0.18**	**0.11**	**0.09**	**0.6**	**0.5**	**34**	**43**	**46**
North America	0.10	0.08	0.07	0.8	0.7	44	48	49
Pac OECD	0.26	0.13	0.10	0.5	0.4	27	39	44
Western Europe	0.20	0.11	0.09	0.6	0.5	32	42	46
OECD	**0.18**	**0.10**	**0.08**	**0.6**	**0.5**	**35**	**44**	**47**
World	**0.13**	**0.09**	**0.07**	**0.7**	**0.6**	**40**	**46**	**48**
Policy Reform								
Africa	0.11	0.10	0.11	0.9	1.0	43	44	42
China+	0.12	0.11	0.12	0.9	1.0	40	42	41
Latin America	0.05	0.07	0.09	1.3	1.7	55	50	46
Middle East	0.11	0.11	0.12	0.9	1.0	41	43	41
South Asia	0.15	0.12	0.13	0.8	0.9	37	40	39
Southeast Asia	0.13	0.11	0.12	0.8	0.9	39	42	41
Developing	**0.12**	**0.11**	**0.12**	**0.9**	**1.0**	**41**	**43**	**41**
Eastern Europe	0.22	0.15	0.15	0.7	0.7	30	36	37
FSU	0.16	0.13	0.13	0.8	0.8	35	39	39
Transitional	**0.18**	**0.13**	**0.13**	**0.8**	**0.8**	**34**	**39**	**39**
North America	0.10	0.10	0.11	1.0	1.1	44	44	42
Pac OECD	0.26	0.17	0.15	0.6	0.6	27	35	36
Western Europe	0.20	0.14	0.14	0.7	0.7	32	38	38
OECD	**0.18**	**0.13**	**0.13**	**0.7**	**0.7**	**35**	**40**	**39**
World	**0.13**	**0.11**	**0.12**	**0.8**	**0.9**	**40**	**42**	**41**

S-2 Hunger

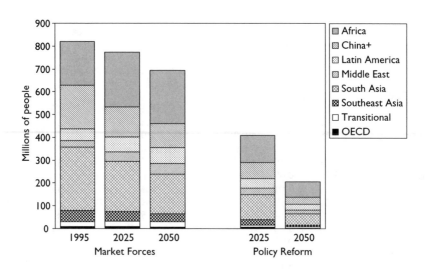

S-2 Hunger

	Incidence (% of population)			Incidence (millions)			Index (1995 = 1)	
	1995	2025	2050	1995	2025	2050	2025	2050
Market Forces								
Africa	27	19	13	192	240	233	1.3	1.2
China+	14	8	6	190	131	105	0.7	0.6
Latin America	11	10	9	53	66	70	1.3	1.3
Middle East	15	13	11	27	43	47	1.6	1.8
South Asia	23	12	8	279	220	175	0.8	0.6
Southeast Asia	11	7	5	49	42	34	0.9	0.7
Developing	**18**	**12**	**9**	**789**	**742**	**664**	**0.9**	**0.8**
Eastern Europe	1	3	4	1	3	3	2.3	2.6
FSU	7	7	7	20	21	20	1.0	1.0
Transitional	**5**	**6**	**6**	**21**	**24**	**23**	**1.1**	**1.1**
North America	2	1	0	7	3	1	0.4	0.2
Pac OECD	0	0	0	0	0	0	1.1	0.8
Western Europe	1	1	1	2	5	6	2.1	2.4
OECD	**1**	**1**	**1**	**10**	**8**	**7**	**0.9**	**0.8**
World	**14**	**10**	**8**	**820**	**774**	**694**	**0.9**	**0.8**
Policy Reform								
Africa	27	9	4	192	119	66	0.6	0.3
China+	14	4	2	190	70	30	0.4	0.2
Latin America	11	6	3	53	42	26	0.8	0.5
Middle East	15	9	4	27	29	18	1.1	0.7
South Asia	23	6	2	279	110	48	0.4	0.2
Southeast Asia	11	4	1	49	22	8	0.4	0.2
Developing	**18**	**6**	**3**	**789**	**393**	**196**	**0.5**	**0.2**
Eastern Europe	1	1	1	1	1	1	0.9	0.5
FSU	7	3	2	20	10	5	0.5	0.2
Transitional	**5**	**3**	**2**	**21**	**11**	**5**	**0.5**	**0.3**
North America	2	1	0	7	3	1	0.5	0.2
Pac OECD	0	0	0	0	0	0	1.0	0.5
Western Europe	1	1	0	2	3	2	1.2	0.8
OECD	**1**	**1**	**0**	**10**	**7**	**3**	**0.7**	**0.3**
World	**14**	**5**	**2**	**820**	**410**	**205**	**0.5**	**0.2**

En-1 Primary energy requirements by region

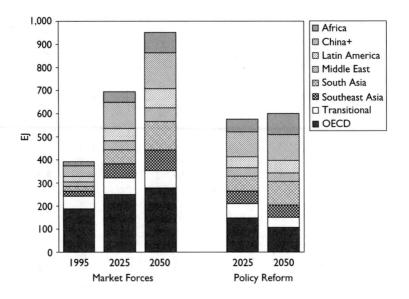

En-1 Primary energy requirements by region

	Primary requirements (EJ)			Growth rate (%/year)			Index (1995 = 1)		Intensity (MJ/$PPP)		
	1995	2025	2050	1995–2025	2025–2050	1995–2050	2025	2050	1995	2025	2050
Market Forces											
Africa	19	46	88	3.0	2.6	2.8	2.4	4.7	14	11	9
China+	45	113	156	3.1	1.3	2.3	2.5	3.4	12	9	7
Latin America	24	53	83	2.7	1.8	2.3	2.2	3.4	9	7	6
Middle East	19	39	59	2.5	1.7	2.1	2.1	3.1	15	10	7
South Asia	21	61	123	3.6	2.9	3.3	2.9	5.9	11	9	8
Southeast Asia	22	61	89	3.5	1.5	2.6	2.8	4.1	9	8	6
Developing	**150**	**373**	**598**	**3.1**	**1.9**	**2.5**	**2.5**	**4.0**	**11**	**9**	**7**
Eastern Europe	11	14	13	0.8	−0.2	0.4	1.3	1.2	19	14	11
FSU	44	58	62	0.9	0.3	0.6	1.3	1.4	40	29	23
Transitional	**55**	**72**	**75**	**0.9**	**0.2**	**0.6**	**1.3**	**1.4**	**33**	**25**	**20**
North America	99	137	161	1.1	0.6	0.9	1.4	1.6	12	8	6
Pac OECD	23	31	32	1.0	0.1	0.6	1.3	1.4	7	5	4
Western Europe	66	81	84	0.7	0.2	0.5	1.2	1.3	8	5	4
OECD	**188**	**249**	**278**	**1.0**	**0.4**	**0.7**	**1.3**	**1.5**	**10**	**6**	**5**
World	**392**	**695**	**951**	**1.9**	**1.3**	**1.6**	**1.8**	**2.4**	**11**	**8**	**6**
Policy Reform											
Africa	19	55	92	3.7	2.0	2.9	2.9	4.9	14	9	6
China+	45	108	112	2.9	0.1	1.7	2.4	2.5	12	8	5
Latin America	24	47	55	2.3	0.6	1.5	2.0	2.3	9	6	4
Middle East	19	37	37	2.3	0.0	1.2	2.0	1.9	15	9	4
South Asia	21	64	102	3.8	1.9	2.9	3.1	4.9	11	8	6
Southeast Asia	22	54	53	3.0	−0.1	1.6	2.5	2.4	9	6	3
Developing	**150**	**366**	**450**	**3.0**	**0.8**	**2.0**	**2.4**	**3.0**	**11**	**8**	**5**
Eastern Europe	11	12	8	0.5	−1.7	−0.5	1.2	0.8	19	12	6
FSU	44	50	36	0.4	−1.3	−0.4	1.1	0.8	40	21	10
Transitional	**55**	**62**	**44**	**0.4**	**−1.4**	**−0.4**	**1.1**	**0.8**	**33**	**18**	**9**
North America	99	76	58	−0.9	−1.1	−1.0	0.8	0.6	12	6	4
Pac OECD	23	19	12	−0.7	−1.8	−1.2	0.8	0.5	7	4	2
Western Europe	66	53	37	−0.7	−1.5	−1.1	0.8	0.6	8	4	2
OECD	**188**	**148**	**107**	**−0.8**	**−1.3**	**−1.0**	**0.8**	**0.6**	**10**	**5**	**3**
World	**392**	**576**	**600**	**1.3**	**0.2**	**0.8**	**1.5**	**1.5**	**11**	**7**	**4**

En-2 Primary energy requirements by source

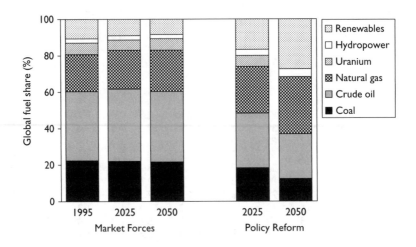

En-2 Primary energy requirements by source

	Primary energy (EJ)			Share of total (%)			Growth rate (%/year)			Index (1995 = 1)	
	1995	2025	2050	1995	2025	2050	1995–2025	2025–2050	1995–2050	2025	2050
Market Forces											
Coal	87	153	207	22	22	22	1.9	1.2	1.6	1.8	2.4
Crude oil	149	276	367	38	40	39	2.1	1.2	1.7	1.8	2.5
Natural gas	79	147	214	20	21	23	2.1	1.5	1.8	1.9	2.7
Uranium	25	40	61	6	6	6	1.5	1.7	1.6	1.6	2.4
Hydropower	9	17	22	2	2	2	2.2	1.1	1.7	1.9	2.5
Renewables	42	62	79	11	9	8	1.3	1.0	1.2	1.5	1.9
Total	**392**	**695**	**951**	**100**	**100**	**100**	**1.9**	**1.3**	**1.6**	**1.8**	**2.4**
Policy Reform											
Coal	87	106	74	22	18	12	0.6	−1.4	−0.3	1.2	0.9
Crude oil	149	173	149	38	30	25	0.5	−0.6	0.0	1.2	1.0
Natural gas	79	148	188	20	26	31	2.1	1.0	1.6	1.9	2.4
Uranium	25	35	—	6	6	—	1.1	—	—	1.4	—
Hydropower	9	18	25	2	3	4	2.4	1.3	1.9	2.0	2.8
Renewables	42	97	165	11	17	27	2.8	2.2	2.5	2.3	3.9
Total	**392**	**576**	**600**	**100**	**100**	**100**	**1.3**	**0.2**	**0.8**	**1.5**	**1.5**

En-3 Final fuel demand by region

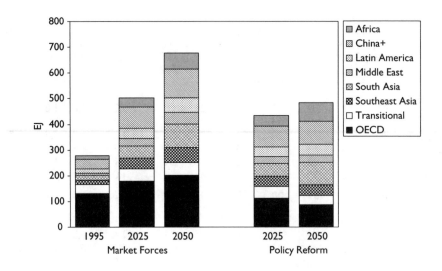

En-3 Final fuel demand by region

	Final requirements (EJ)			Growth rate (%/year)			Index (1995 = 1)		Intensity (MJ/$PPP)		
	1995	2025	2050	1995–2025	2025–2050	1995–2050	2025	2050	1995	2025	2050
Market Forces											
Africa	14	35	64	3.1	2.4	2.8	2.5	4.5	10	8	7
China+	38	84	111	2.7	1.2	2.0	2.2	3.0	10	7	5
Latin America	17	40	57	2.9	1.4	2.2	2.3	3.3	6	5	4
Middle East	9	28	44	3.8	1.8	2.9	3.0	4.7	7	7	5
South Asia	17	47	93	3.4	2.7	3.1	2.7	5.3	9	7	6
Southeast Asia	16	41	57	3.2	1.3	2.3	2.5	3.5	6	5	4
Developing	111	275	426	3.1	1.8	2.5	2.5	3.8	8	7	5
Eastern Europe	7	9	8	0.9	−0.2	0.4	1.3	1.3	12	9	7
FSU	29	40	43	1.1	0.3	0.7	1.4	1.5	27	20	16
Transitional	36	49	52	1.0	0.2	0.6	1.4	1.4	22	17	13
North America	66	98	117	1.3	0.7	1.0	1.5	1.8	8	5	4
Pac OECD	17	21	22	0.7	0.1	0.5	1.2	1.3	5	3	3
Western Europe	47	59	62	0.7	0.2	0.5	1.2	1.3	6	4	3
OECD	131	178	201	1.0	0.5	0.8	1.4	1.5	7	4	3
World	279	503	678	2.0	1.2	1.6	1.8	2.4	8	6	5
Policy Reform											
Africa	14	42	73	3.7	2.2	3.0	3.0	5.2	10	7	5
China+	38	81	89	2.6	0.4	1.6	2.1	2.4	10	6	4
Latin America	17	38	41	2.7	0.3	1.6	2.2	2.4	6	5	3
Middle East	9	25	29	3.4	0.6	2.1	2.7	3.1	7	6	4
South Asia	17	50	87	3.6	2.2	3.0	2.9	5.0	9	6	5
Southeast Asia	16	39	42	3.0	0.2	1.7	2.4	2.6	6	5	2
Developing	111	276	361	3.1	1.1	2.2	2.5	3.2	8	6	4
Eastern Europe	7	9	6	0.8	−1.2	−0.2	1.3	0.9	12	8	5
FSU	29	38	30	0.9	−0.9	0.0	1.3	1.0	27	16	9
Transitional	36	47	36	0.9	−1.0	0.0	1.3	1.0	22	14	8
North America	66	59	48	−0.4	−0.8	−0.6	0.9	0.7	8	4	3
Pac OECD	17	13	9	−0.8	−1.4	−1.1	0.8	0.5	5	3	2
Western Europe	47	40	29	−0.6	−1.2	−0.9	0.8	0.6	6	3	2
OECD	131	112	87	−0.5	−1.0	−0.7	0.9	0.7	7	4	2
World	279	434	484	1.5	0.4	1.0	1.6	1.7	8	5	3

En-4 Final fuel demand by sector

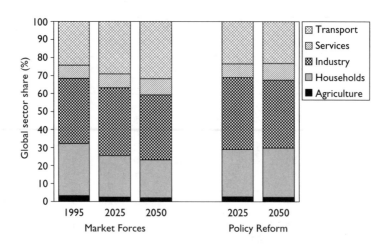

En-4 Final fuel demand by sector

	Final energy demand (EJ)			Share of total (%)			Growth rate (%/year)			Index (1995 = 1)	
	1995	2025	2050	1995	2025	2050	1995–2025	2025–2050	1995–2050	2025	2050
Market Forces											
Agriculture	9	12	14	3	2	2	1.0	0.6	0.8	1.3	1.6
Households	80	116	143	29	23	21	1.2	0.8	1.0	1.4	1.8
Industry	101	190	245	36	38	36	2.1	1.0	1.6	1.9	2.4
Services	20	39	61	7	8	9	2.2	1.8	2.0	1.9	3.0
Transport	68	146	215	24	29	32	2.6	1.6	2.1	2.2	3.2
Total	**279**	**503**	**678**	**100**	**100**	**100**	**2.0**	**1.2**	**1.6**	**1.8**	**2.4**
Policy Reform											
Agriculture	9	12	12	3	3	2	0.8	0.1	0.5	1.3	1.3
Households	80	115	132	29	26	27	1.2	0.6	0.9	1.4	1.6
Industry	101	173	183	36	40	38	1.8	0.2	1.1	1.7	1.8
Services	20	33	44	7	8	9	1.6	1.2	1.4	1.6	2.2
Transport	68	102	113	24	23	23	1.4	0.4	0.9	1.5	1.7
Total	**279**	**434**	**484**	**100**	**100**	**100**	**1.5**	**0.4**	**1.0**	**1.6**	**1.7**

En-5 Fossil-fuel reserves

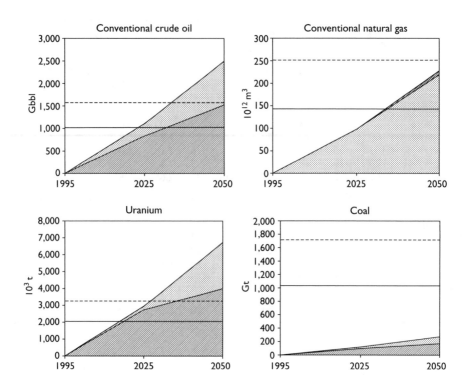

En-5 Fossil-fuel reserves

	Coal $10^9 t$	Crude oil $10^9 bbls$	Natural gas $10^{12} m^3$	Uranium $10^3 t$	Share of reserves (%)			
					Coal	Crude oil	Natural gas	Uranium
Total remaining reserves (Proven + undiscovered, 1995)								
Africa	62	84	15	1,142	4	5	6	35
China+	265	276	4	5	15	17	2	0
Latin America	20	407	20	174	1	26	8	5
Middle East	0	670	80	0	0	42	31	0
South Asia	117	7	3	13	7	0	1	0
Southeast Asia	34	17	9	44	2	1	4	1
Developing	**499**	**1,461**	**131**	**1,378**	**29**	**92**	**52**	**42**
Eastern Europe	92	2	2	21	5	0	1	1
FSU	241	58	91	0	14	4	36	0
Transitional	**333**	**60**	**92**	**21**	**19**	**4**	**36**	**1**
North America	249	45	21	799	15	3	8	24
Pac OECD	525	3	2	926	31	0	1	28
Western Europe	105	25	8	157	6	2	3	5
OECD	**879**	**73**	**31**	**1,882**	**51**	**5**	**12**	**57**
World	**1,711**	**1,594**	**254**	**3,281**	**100**	**100**	**100**	**100**
World (EJ)	**51,331**	**9,120**	**8,905**	**1,102**				

F-1 Diets

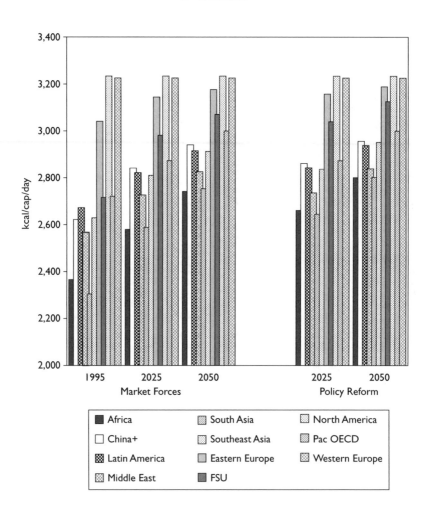

F-1 Diets

	Avg. daily consumption (kcal per capita)			Share from animal products (%)		
	1995	2025	2050	1995	2025	2050
Market Forces						
Africa	2,370	2,580	2,740	7	8	10
China+	2,620	2,840	2,940	16	19	21
Latin America	2,670	2,820	2,920	17	19	21
Middle East	2,570	2,730	2,830	11	12	14
South Asia	2,300	2,590	2,750	7	10	12
Southeast Asia	2,630	2,810	2,910	8	11	13
Developing	**2,500**	**2,710**	**2,830**	**11**	**13**	**15**
Eastern Europe	3,040	3,140	3,180	25	26	26
FSU	2,720	2,980	3,070	23	24	25
Transitional	**2,800**	**3,020**	**3,090**	**23**	**24**	**25**
North America	3,230	3,230	3,230	32	31	30
Pac OECD	2,720	2,870	3,000	24	25	26
Western Europe	3,230	3,230	3,230	29	29	29
OECD	**3,150**	**3,180**	**3,200**	**29**	**29**	**29**
World	**2,620**	**2,780**	**2,880**	**16**	**16**	**17**
Policy Reform						
Africa	2,370	2,660	2,800	7	10	12
China+	2,620	2,860	2,960	16	19	21
Latin America	2,670	2,840	2,940	17	20	21
Middle East	2,570	2,740	2,840	11	12	14
South Asia	2,300	2,650	2,800	7	11	13
Southeast Asia	2,630	2,840	2,950	8	11	14
Developing	**2,500**	**2,750**	**2,870**	**11**	**14**	**16**
Eastern Europe	3,040	3,160	3,190	25	26	27
FSU	2,720	3,040	3,120	23	24	25
Transitional	**2,800**	**3,070**	**3,140**	**23**	**25**	**26**
North America	3,230	3,230	3,230	32	31	30
Pac OECD	2,720	2,870	3,000	24	24	25
Western Europe	3,230	3,230	3,230	29	29	29
OECD	**3,150**	**3,180**	**3,200**	**29**	**29**	**29**
World	**2,620**	**2,820**	**2,920**	**16**	**17**	**18**

F-2 Meat and milk requirements and production

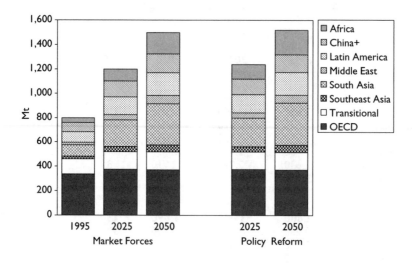

F-2 Meat and milk requirements and production

	Requirements (Mt)			Production (Mt)			Self-sufficiency ratio			Feedlot production (%)		
	1995	2025	2050	1995	2025	2050	1995	2025	2050	1995	2025	2050
Market Forces												
Africa	40	101	175	34	78	128	0.8	0.8	0.7	5	13	17
China+	79	130	151	77	127	152	1.0	1.0	1.0	30	44	52
Latin America	85	146	190	79	147	190	0.9	1.0	1.0	12	21	20
Middle East	18	42	65	14	32	47	0.8	0.8	0.7	31	34	36
South Asia	95	219	338	96	205	306	1.0	0.9	0.9	1	10	19
Southeast Asia	20	41	59	14	28	39	0.7	0.7	0.7	27	41	49
Developing	**337**	**679**	**977**	**314**	**617**	**862**	**0.9**	**0.9**	**0.9**	**14**	**24**	**28**
Eastern Europe	32	33	30	35	38	39	1.1	1.1	1.3	50	54	55
FSU	95	113	115	94	121	139	1.0	1.1	1.2	27	34	36
Transitional	**127**	**146**	**145**	**129**	**159**	**178**	**1.0**	**1.1**	**1.2**	**32**	**38**	**40**
North America	124	147	154	124	146	153	1.0	1.0	1.0	43	47	49
Pac OECD	30	35	35	38	46	50	1.3	1.3	1.4	13	17	18
Western Europe	181	193	184	199	227	242	1.1	1.2	1.3	27	30	31
OECD	**336**	**375**	**373**	**361**	**419**	**444**	**1.1**	**1.1**	**1.2**	**30**	**34**	**36**
World	**799**	**1,200**	**1,496**	**805**	**1,195**	**1,484**	**1.0**	**1.0**	**1.0**	**21**	**28**	**31**
Policy Reform												
Africa	40	118	200	34	90	145	0.8	0.8	0.7	5	15	18
China+	79	131	147	77	129	148	1.0	1.0	1.0	30	45	51
Latin America	85	147	186	79	148	188	0.9	1.0	1.0	12	29	36
Middle East	18	42	63	14	31	45	0.8	0.8	0.7	31	34	36
South Asia	95	235	348	96	218	314	1.0	0.9	0.9	1	14	22
Southeast Asia	20	43	62	14	29	41	0.7	0.7	0.7	27	42	51
Developing	**337**	**716**	**1,004**	**314**	**646**	**881**	**0.9**	**0.9**	**0.9**	**14**	**27**	**33**
Eastern Europe	32	33	29	35	38	38	1.1	1.2	1.3	50	54	56
FSU	95	115	114	94	125	141	1.0	1.1	1.2	27	35	38
Transitional	**127**	**148**	**143**	**129**	**164**	**180**	**1.0**	**1.1**	**1.3**	**32**	**39**	**41**
North America	124	147	154	124	146	153	1.0	1.0	1.0	43	47	49
Pac OECD	30	34	34	38	45	48	1.3	1.3	1.4	13	17	17
Western Europe	181	192	183	199	230	245	1.1	1.2	1.3	27	30	31
OECD	**336**	**374**	**371**	**361**	**422**	**447**	**1.1**	**1.1**	**1.2**	**30**	**34**	**35**
World	**799**	**1,237**	**1,519**	**805**	**1,232**	**1,507**	**1.0**	**1.0**	**1.0**	**21**	**30**	**34**

F-3 Fish requirements and production

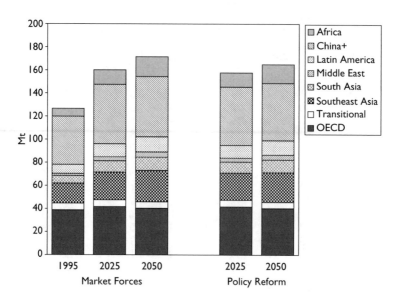

F-3 Fish requirements and production

	Requirements (Mt)			Production (Mt)			Self-sufficiency ratio		
	1995	2025	2050	1995	2025	2050	1995	2025	2050
Market Forces									
Africa	7	13	17	6	10	13	0.8	0.8	0.7
China+	42	52	52	37	45	45	0.9	0.9	0.9
Latin America	8	11	13	22	33	38	2.8	3.0	2.9
Middle East	2	3	5	1	1	2	0.4	0.4	0.3
South Asia	7	10	12	7	12	14	1.1	1.3	1.2
Southeast Asia	17	24	27	16	22	24	0.9	0.9	0.9
Developing	**82**	**113**	**126**	**88**	**123**	**136**	**1.1**	**1.1**	**1.1**
Eastern Europe	2	1	1	1	1	0	0.4	0.4	0.4
FSU	4	4	4	5	6	6	1.2	1.4	1.4
Transitional	**6**	**6**	**6**	**6**	**7**	**6**	**1.0**	**1.2**	**1.1**
North America	8	9	10	7	8	8	0.9	0.8	0.8
Pac OECD	14	15	13	8	9	8	0.6	0.6	0.6
Western Europe	17	18	17	13	14	13	0.8	0.8	0.8
OECD	**39**	**42**	**40**	**28**	**30**	**30**	**0.7**	**0.7**	**0.7**
World	**127**	**160**	**172**	**122**	**160**	**172**	**1.0**	**1.0**	**1.0**
Policy Reform									
Africa	7	12	16	6	10	12	0.8	0.8	0.7
China+	42	50	50	37	44	43	0.9	0.9	0.9
Latin America	8	11	12	22	33	36	2.8	3.0	2.9
Middle East	2	3	4	1	1	2	0.4	0.4	0.3
South Asia	7	10	11	7	12	13	1.1	1.3	1.2
Southeast Asia	17	23	26	16	21	23	0.9	0.9	0.9
Developing	**82**	**110**	**119**	**88**	**121**	**129**	**1.1**	**1.1**	**1.1**
Eastern Europe	2	1	1	1	1	0	0.4	0.4	0.4
FSU	4	4	4	5	6	6	1.2	1.4	1.4
Transitional	**6**	**6**	**5**	**6**	**7**	**6**	**1.0**	**1.2**	**1.1**
North America	8	9	10	7	8	8	0.9	0.8	0.8
Pac OECD	14	15	13	8	9	8	0.6	0.6	0.6
Western Europe	17	18	17	13	14	13	0.8	0.8	0.8
OECD	**39**	**42**	**40**	**28**	**30**	**30**	**0.7**	**0.7**	**0.7**
World	**127**	**158**	**165**	**122**	**158**	**165**	**1.0**	**1.0**	**1.0**

F-4 Crop requirements and production

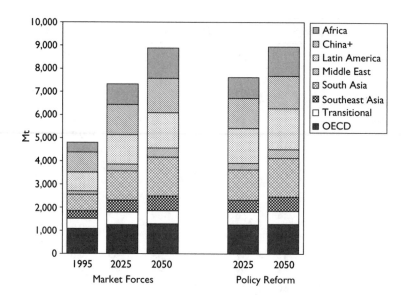

F-4 Crop requirements and production

	Requirements (Mt)			Production (Mt)			Net exports (Mt)			Self-sufficiency ratio		
	1995	2025	2050	1995	2025	2050	1995	2025	2050	1995	2025	2050
Market Forces												
Africa	421	892	1,310	343	671	947	−78	−221	−363	0.8	0.8	0.7
China+	867	1,303	1,484	834	1,201	1,212	−33	−102	−272	1.0	0.9	0.8
Latin America	814	1,277	1,520	837	1,413	1,969	23	136	449	1.0	1.1	1.3
Middle East	147	290	394	88	166	217	−59	−124	−176	0.6	0.6	0.6
South Asia	706	1,259	1,683	737	1,202	1,397	31	−57	−286	1.0	1.0	0.8
Southeast Asia	329	518	632	396	624	779	67	105	146	1.2	1.2	1.2
Developing	**3,284**	**5,540**	**7,023**	**3,235**	**5,277**	**6,521**	**−50**	**−263**	**−502**	**1.0**	**1.0**	**0.9**
Eastern Europe	132	137	130	135	157	168	3	20	39	1.0	1.1	1.3
FSU	302	400	436	244	433	529	−58	33	92	0.8	1.1	1.2
Transitional	**434**	**537**	**566**	**379**	**590**	**697**	**−55**	**53**	**131**	**0.9**	**1.1**	**1.2**
North America	430	531	574	537	672	767	107	142	193	1.2	1.3	1.3
Pac OECD	120	137	136	111	149	187	−9	12	52	0.9	1.1	1.4
Western Europe	536	587	585	539	643	712	3	56	127	1.0	1.1	1.2
OECD	**1,086**	**1,255**	**1,295**	**1,187**	**1,465**	**1,667**	**101**	**210**	**371**	**1.1**	**1.2**	**1.3**
World	**4,804**	**7,331**	**8,885**	**4,800**	**7,331**	**8,885**	**−4**	**0**	**0**	**1.0**	**1.0**	**1.0**
Policy Reform												
Africa	421	910	1,268	343	632	824	−78	−278	−444	0.8	0.7	0.7
China+	867	1,295	1,405	834	1,269	1,392	−33	−27	−13	1.0	1.0	1.0
Latin America	814	1,510	1,761	837	1,610	1,924	23	100	163	1.0	1.1	1.1
Middle East	147	282	372	88	139	172	−59	−143	−200	0.6	0.5	0.5
South Asia	706	1,303	1,680	737	1,203	1,438	31	−100	−242	1.0	0.9	0.9
Southeast Asia	329	515	616	396	611	704	67	96	89	1.2	1.2	1.1
Developing	**3,284**	**5,816**	**7,100**	**3,235**	**5,465**	**6,454**	**−50**	**−351**	**−646**	**1.0**	**0.9**	**0.9**
Eastern Europe	132	137	125	135	158	159	3	21	35	1.0	1.2	1.3
FSU	302	413	435	244	459	532	−58	45	97	0.8	1.1	1.2
Transitional	**434**	**550**	**560**	**379**	**617**	**691**	**−55**	**67**	**131**	**0.9**	**1.1**	**1.2**
North America	430	532	577	537	734	849	107	202	272	1.2	1.4	1.5
Pac OECD	120	138	133	111	148	158	−9	11	25	0.9	1.1	1.2
Western Europe	536	591	580	539	663	796	3	72	217	1.0	1.1	1.4
OECD	**1,086**	**1,261**	**1,289**	**1,187**	**1,545**	**1,803**	**101**	**285**	**514**	**1.1**	**1.2**	**1.4**
World	**4,804**	**7,627**	**8,949**	**4,800**	**7,627**	**8,949**	**−4**	**0**	**0**	**1.0**	**1.0**	**1.0**

F-5 Cereal yields

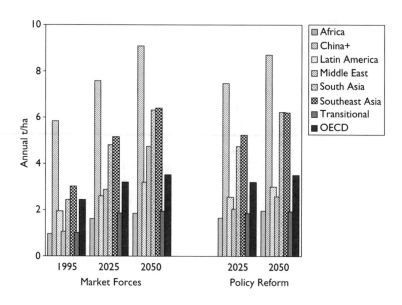

F-5 Cereal yields

	Harvest yield (t/ha)			Cropping intensity (harvests/year)			Annual yield (t/ha/year)			Index (1995 = 1)	
	1995	2025	2050	1995	2025	2050	1995	2025	2050	2025	2050
Market Forces											
Africa	1.1	1.7	1.9	0.88	0.95	1.00	1.0	1.6	1.9	1.7	1.9
China+	4.5	5.6	6.4	1.29	1.35	1.42	5.8	7.6	9.1	1.3	1.6
Latin America	2.5	3.2	3.7	0.77	0.82	0.87	2.0	2.6	3.2	1.3	1.6
Middle East	1.6	3.1	4.1	0.65	0.92	1.15	1.1	2.9	4.8	2.7	4.5
South Asia	2.2	3.8	4.7	1.13	1.25	1.33	2.4	4.8	6.3	2.0	2.6
Southeast Asia	3.1	5.0	6.0	0.98	1.03	1.07	3.0	5.2	6.4	1.7	2.1
Developing	**2.6**	**3.7**	**4.1**	**1.00**	**1.06**	**1.11**	**2.6**	**4.0**	**4.5**	**1.5**	**1.7**
Eastern Europe	3.3	5.0	5.5	0.80	0.80	0.80	2.6	4.0	4.4	1.5	1.7
FSU	1.3	2.8	3.0	0.57	0.57	0.57	0.7	1.6	1.7	2.2	2.3
Transitional	**1.7**	**3.1**	**3.3**	**0.60**	**0.59**	**0.59**	**1.0**	**1.9**	**1.9**	**1.8**	**1.9**
North America	4.2	5.1	5.3	0.60	0.64	0.66	2.5	3.2	3.6	1.3	1.4
Pac OECD	2.5	3.3	3.8	0.42	0.47	0.51	1.0	1.5	1.9	1.5	1.8
Western Europe	4.2	5.4	5.7	0.75	0.75	0.75	3.1	4.0	4.3	1.3	1.4
OECD	**4.0**	**5.0**	**5.3**	**0.62**	**0.64**	**0.67**	**2.4**	**3.2**	**3.5**	**1.3**	**1.4**
World	**2.7**	**3.9**	**4.2**	**0.80**	**0.84**	**0.85**	**2.2**	**3.3**	**3.6**	**1.5**	**1.6**
Policy Reform											
Africa	1.1	1.7	2.0	0.88	0.95	1.00	1.0	1.7	2.0	1.7	2.0
China+	4.5	5.6	6.2	1.29	1.34	1.39	5.8	7.5	8.7	1.3	1.5
Latin America	2.5	3.1	3.5	0.77	0.82	0.87	2.0	2.6	3.0	1.3	1.5
Middle East	1.6	2.7	3.4	0.65	0.74	0.76	1.1	2.0	2.6	1.9	2.4
South Asia	2.2	3.8	4.7	1.13	1.25	1.33	2.4	4.8	6.2	1.9	2.5
Southeast Asia	3.1	5.1	5.9	0.98	1.03	1.06	3.0	5.2	6.2	1.7	2.1
Developing	**2.6**	**3.7**	**4.2**	**1.00**	**1.05**	**1.09**	**2.6**	**3.9**	**4.6**	**1.5**	**1.8**
Eastern Europe	3.3	5.0	5.5	0.80	0.80	0.80	2.6	4.0	4.4	1.5	1.7
FSU	1.3	2.8	3.0	0.57	0.57	0.57	0.7	1.6	1.7	2.2	2.3
Transitional	**1.7**	**3.1**	**3.3**	**0.60**	**0.59**	**0.59**	**1.0**	**1.9**	**1.9**	**1.8**	**1.9**
North America	4.2	5.1	5.3	0.60	0.64	0.66	2.5	3.2	3.5	1.3	1.4
Pac OECD	2.5	3.2	3.5	0.42	0.47	0.51	1.0	1.5	1.8	1.5	1.7
Western Europe	4.2	5.4	5.7	0.75	0.75	0.75	3.1	4.0	4.3	1.3	1.4
OECD	**4.0**	**5.0**	**5.2**	**0.62**	**0.64**	**0.67**	**2.4**	**3.2**	**3.5**	**1.3**	**1.4**
World	**2.7**	**3.9**	**4.3**	**0.80**	**0.83**	**0.85**	**2.2**	**3.2**	**3.6**	**1.5**	**1.6**

F-6 Cropland

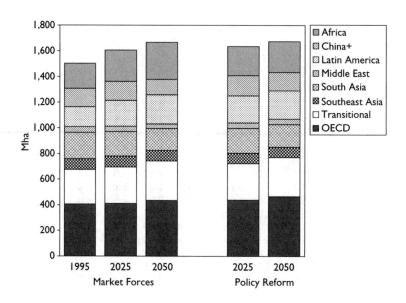

F-6 Cropland

	Cropland (Mha)			Growth rate (%/year)			Index (1995=1)	
	1995	2025	2050	1995–2025	2025–2050	1995–2050	2025	2050
Market Forces								
Africa	195	244	289	0.7	0.7	0.7	1.2	1.5
China+	144	149	121	0.1	−0.8	−0.3	1.0	0.8
Latin America	155	202	228	0.9	0.5	0.7	1.3	1.5
Middle East	45	40	35	−0.4	−0.5	−0.4	0.9	0.8
South Asia	205	191	170	−0.2	−0.5	−0.3	0.9	0.8
Southeast Asia	84	86	83	0.1	−0.2	0.0	1.0	1.0
Developing	**829**	**912**	**925**	**0.3**	**0.1**	**0.2**	**1.1**	**1.1**
Eastern Europe	40	34	32	−0.5	−0.3	−0.4	0.9	0.8
FSU	229	249	275	0.3	0.4	0.3	1.1	1.2
Transitional	**268**	**283**	**307**	**0.2**	**0.3**	**0.2**	**1.1**	**1.1**
North America	227	229	237	0.0	0.1	0.1	1.0	1.0
Pac OECD	55	56	62	0.0	0.4	0.2	1.0	1.1
Western Europe	124	125	135	0.0	0.3	0.2	1.0	1.1
OECD	**406**	**411**	**434**	**0.0**	**0.2**	**0.1**	**1.0**	**1.1**
World	**1,503**	**1,606**	**1,667**	**0.2**	**0.1**	**0.2**	**1.1**	**1.1**
Policy Reform								
Africa	195	227	240	0.5	0.2	0.4	1.2	1.2
China+	144	159	144	0.3	−0.4	0.0	1.1	1.0
Latin America	155	211	221	1.0	0.2	0.6	1.4	1.4
Middle East	45	42	43	−0.2	0.0	−0.1	0.9	0.9
South Asia	205	192	176	−0.2	−0.3	−0.3	0.9	0.9
Southeast Asia	84	83	80	−0.1	−0.1	−0.1	1.0	0.9
Developing	**829**	**913**	**903**	**0.3**	**0.0**	**0.2**	**1.1**	**1.1**
Eastern Europe	40	33	30	−0.6	−0.3	−0.5	0.8	0.8
FSU	229	251	275	0.3	0.4	0.3	1.1	1.2
Transitional	**268**	**284**	**305**	**0.2**	**0.3**	**0.2**	**1.1**	**1.1**
North America	227	253	268	0.4	0.2	0.3	1.1	1.2
Pac OECD	55	56	52	0.0	−0.3	−0.1	1.0	0.9
Western Europe	124	129	145	0.1	0.5	0.3	1.0	1.2
OECD	**406**	**438**	**466**	**0.3**	**0.2**	**0.3**	**1.1**	**1.1**
World	**1,503**	**1,635**	**1,673**	**0.3**	**0.1**	**0.2**	**1.1**	**1.1**

F-7 Irrigation

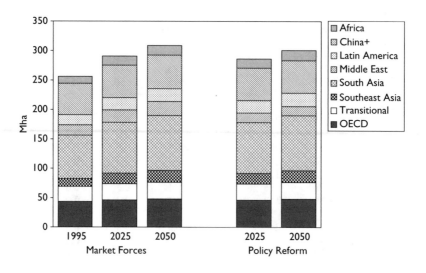

F-7 Irrigation

	Irrigated cropland (Mha)			Index (1995 = 1)		Fraction of cropland irrigated (%)		
	1995	2025	2050	2025	2050	1995	2025	2050
Market Forces								
Africa	12	15	17	1.2	1.4	6	6	6
China+	53	56	57	1.0	1.1	37	37	47
Latin America	18	21	23	1.2	1.3	11	10	10
Middle East	17	21	23	1.2	1.3	39	53	65
South Asia	74	87	94	1.2	1.3	36	45	55
Southeast Asia	14	17	19	1.2	1.4	17	20	23
Developing	**188**	**217**	**233**	**1.2**	**1.2**	**23**	**24**	**25**
Eastern Europe	5	5	6	1.1	1.2	12	15	17
FSU	20	22	23	1.1	1.2	9	9	8
Transitional	**25**	**28**	**29**	**1.1**	**1.2**	**9**	**10**	**9**
North America	22	23	24	1.1	1.1	10	10	10
Pac OECD	5	5	5	1.0	1.0	10	10	9
Western Europe	16	18	19	1.1	1.2	13	14	14
OECD	**44**	**46**	**48**	**1.1**	**1.1**	**11**	**11**	**11**
World	**256**	**291**	**309**	**1.1**	**1.2**	**17**	**18**	**19**
Policy Reform								
Africa	12	15	17	1.2	1.4	6	7	7
China+	53	55	56	1.0	1.0	37	35	39
Latin America	18	21	23	1.2	1.3	11	10	10
Middle East	17	16	16	1.0	0.9	39	39	37
South Asia	74	87	94	1.2	1.3	36	45	53
Southeast Asia	14	17	19	1.2	1.4	17	21	24
Developing	**188**	**212**	**224**	**1.1**	**1.2**	**23**	**23**	**25**
Eastern Europe	5	5	6	1.1	1.2	12	16	18
FSU	20	22	23	1.1	1.2	9	9	8
Transitional	**25**	**28**	**29**	**1.1**	**1.2**	**9**	**10**	**9**
North America	22	23	24	1.1	1.1	10	9	9
Pac OECD	5	5	5	1.0	1.0	10	10	10
Western Europe	16	18	19	1.1	1.2	13	14	13
OECD	**44**	**46**	**48**	**1.1**	**1.1**	**11**	**11**	**10**
World	**256**	**286**	**301**	**1.1**	**1.2**	**17**	**17**	**18**

F-8 Potential cultivable land

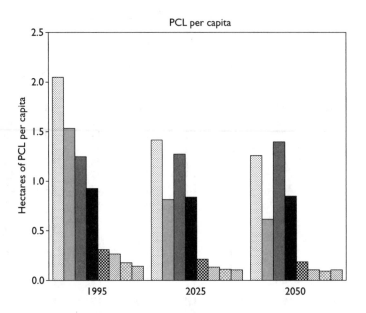

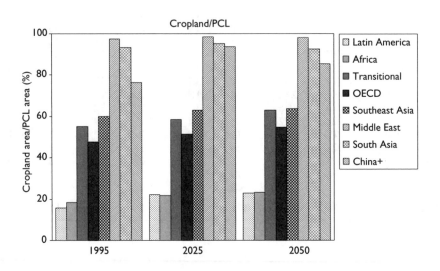

F-8 Potential cultivable land

	PCL (Mha)			PCL per capita (ha per capita)			Cropland/PCL (%)		
	1995	2025	2050	1995	2025	2050	1995	2025	2050
Market Forces									
Africa	1,069	1,014	966	1.5	0.8	0.5	18	24	30
China+	189	158	131	0.1	0.1	0.1	76	94	93
Latin America	975	945	914	2.0	1.4	1.1	16	21	25
Middle East	46	41	36	0.3	0.1	0.1	97	99	98
South Asia	220	198	177	0.2	0.1	0.1	93	96	96
Southeast Asia	140	129	114	0.3	0.2	0.2	60	67	73
Developing	**2,640**	**2,484**	**2,337**	**0.6**	**0.4**	**0.3**	**31**	**37**	**40**
Eastern Europe	64	62	61	0.6	0.7	0.7	62	56	53
FSU	424	421	419	1.5	1.4	1.5	54	59	66
Transitional	**487**	**483**	**480**	**1.2**	**1.2**	**1.3**	**55**	**59**	**64**
North America	480	478	475	1.6	1.3	1.2	47	48	50
Pac OECD	167	166	165	1.1	1.1	1.2	33	34	38
Western Europe	204	202	202	0.4	0.4	0.4	61	62	67
OECD	**851**	**846**	**842**	**0.9**	**0.8**	**0.8**	**48**	**49**	**52**
World	**3,978**	**3,813**	**3,659**	**0.7**	**0.5**	**0.4**	**38**	**42**	**46**
Policy Reform									
Africa	1,069	1,037	1,035	1.5	0.8	0.6	18	22	23
China+	189	170	168	0.1	0.1	0.1	76	93	85
Latin America	975	960	961	2.0	1.4	1.3	16	22	23
Middle East	46	43	43	0.3	0.1	0.1	97	98	98
South Asia	220	202	191	0.2	0.1	0.1	93	95	92
Southeast Asia	140	132	126	0.3	0.2	0.2	60	63	64
Developing	**2,640**	**2,544**	**2,525**	**0.6**	**0.4**	**0.4**	**31**	**36**	**36**
Eastern Europe	64	63	64	0.6	0.7	0.8	62	52	48
FSU	424	422	423	1.5	1.5	1.6	54	59	65
Transitional	**487**	**485**	**486**	**1.2**	**1.3**	**1.4**	**55**	**58**	**63**
North America	480	482	482	1.6	1.3	1.2	47	53	56
Pac OECD	167	167	168	1.1	1.1	1.2	33	33	31
Western Europe	204	204	204	0.4	0.4	0.4	61	63	71
OECD	**851**	**852**	**854**	**0.9**	**0.8**	**0.8**	**48**	**51**	**55**
World	**3,978**	**3,882**	**3,865**	**0.7**	**0.5**	**0.5**	**38**	**42**	**43**

P-1 Water use by region

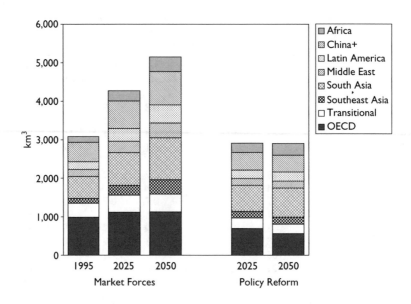

P-1 Water use by region

	Water use (km³)			Growth rate (%/year)			Index (1995 = 1)	
	1995	2025	2050	1995–2025	2025–2050	1995–2050	2025	2050
Market Forces								
Africa	152	268	382	1.9	1.4	1.7	1.8	2.5
China+	506	711	870	1.1	0.8	1.0	1.4	1.7
Latin America	201	331	467	1.7	1.4	1.5	1.6	2.3
Middle East	182	294	385	1.6	1.1	1.4	1.6	2.1
South Asia	569	855	1,092	1.4	1.0	1.2	1.5	1.9
Southeast Asia	124	251	371	2.4	1.6	2.0	2.0	3.0
Developing	**1,733**	**2,710**	**3,566**	**1.5**	**1.1**	**1.3**	**1.6**	**2.1**
Eastern Europe	64	77	76	0.6	0.0	0.3	1.2	1.2
FSU	301	369	384	0.7	0.2	0.4	1.2	1.3
Transitional	**365**	**446**	**460**	**0.7**	**0.1**	**0.4**	**1.2**	**1.3**
North America	588	675	694	0.5	0.1	0.3	1.1	1.2
Pac OECD	99	109	109	0.3	0.0	0.2	1.1	1.1
Western Europe	299	332	324	0.3	−0.1	0.1	1.1	1.1
OECD	**986**	**1,115**	**1,127**	**0.4**	**0.0**	**0.2**	**1.1**	**1.1**
World	**3,084**	**4,271**	**5,154**	**1.1**	**0.8**	**0.9**	**1.4**	**1.7**
Policy Reform								
Africa	152	241	306	1.6	1.0	1.3	1.6	2.0
China+	506	462	438	−0.3	−0.2	−0.3	0.9	0.9
Latin America	201	219	238	0.3	0.3	0.3	1.1	1.2
Middle East	182	177	178	−0.1	0.0	0.0	1.0	1.0
South Asia	569	677	756	0.6	0.4	0.5	1.2	1.3
Southeast Asia	124	168	182	1.0	0.3	0.7	1.4	1.5
Developing	**1,733**	**1,944**	**2,097**	**0.4**	**0.3**	**0.3**	**1.1**	**1.2**
Eastern Europe	64	49	43	−0.9	−0.5	−0.7	0.8	0.7
FSU	301	226	206	−0.9	−0.4	−0.7	0.8	0.7
Transitional	**365**	**275**	**250**	**−0.9**	**−0.4**	**−0.7**	**0.8**	**0.7**
North America	588	424	325	−1.1	−1.1	−1.1	0.7	0.6
Pac OECD	99	90	82	−0.3	−0.4	−0.3	0.9	0.8
Western Europe	299	179	150	−1.7	−0.7	−1.2	0.6	0.5
OECD	**986**	**693**	**558**	**−1.2**	**−0.9**	**−1.0**	**0.7**	**0.6**
World	**3,084**	**2,913**	**2,905**	**−0.2**	**0.0**	**−0.1**	**0.9**	**0.9**

P-2 Water use by sector

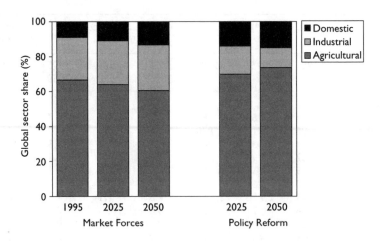

P-2 Water use by sector

	Water use (km^3)			Growth rate (%/year)			Index (1995 = 1)		Sectoral share (%)		
	1995	2025	2050	1995–2025	2025–2050	1995–2050	2025	2050	1995	2025	2050
Market Forces											
Agricultural	2,056	2,732	3,119	1.0	0.5	0.8	1.3	1.5	67	64	61
Industrial	750	1,071	1,349	1.2	0.9	1.1	1.4	1.8	24	25	26
Domestic	278	468	686	1.8	1.5	1.7	1.7	2.5	9	11	13
Total	**3,084**	**4,271**	**5,154**	**1.1**	**0.8**	**0.9**	**1.4**	**1.7**	**100**	**100**	**100**
Policy Reform											
Agricultural	2,056	2,037	2,138	0.0	0.2	0.1	1.0	1.0	67	70	74
Industrial	750	467	330	−1.6	−1.4	−1.5	0.6	0.4	24	16	11
Domestic	278	409	437	1.3	0.3	0.8	1.5	1.6	9	14	15
Total	**3,084**	**2,913**	**2,905**	**−0.2**	**0.0**	**−0.1**	**0.9**	**0.9**	**100**	**100**	**100**

P-3 Water stress

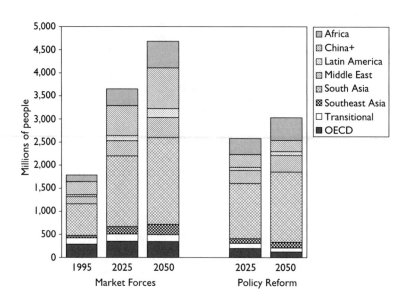

P-3 Water stress

	Use-to-resource ratio			Water stress (million people)			Index (1995 = 1)		Water stress (% population)		
	1995	2025	2050	1995	2025	2050	2025	2050	1995	2025	2050
Market Forces											
Africa	0.04	0.07	0.10	144	369	574	2.6	4.0	21	28	33
China+	0.16	0.23	0.28	284	646	885	2.3	3.1	21	39	53
Latin America	0.02	0.03	0.04	41	112	193	2.7	4.7	9	16	24
Middle East	0.62	1.00	1.31	154	327	433	2.1	2.8	89	100	100
South Asia	0.26	0.40	0.51	688	1,530	1,879	2.2	2.7	56	83	87
Southeast Asia	0.02	0.05	0.07	52	165	226	3.2	4.3	11	26	32
Developing	**0.07**	**0.11**	**0.14**	**1,363**	**3,149**	**4,190**	**2.3**	**3.1**	**31**	**49**	**56**
Eastern Europe	0.31	0.38	0.37	58	65	57	1.1	1.0	59	69	69
FSU	0.07	0.08	0.08	79	92	91	1.2	1.2	27	31	32
Transitional	**0.08**	**0.09**	**0.10**	**137**	**157**	**148**	**1.1**	**1.1**	**35**	**40**	**40**
North America	0.11	0.13	0.13	89	121	133	1.4	1.5	30	33	34
Pac OECD	0.08	0.08	0.08	20	27	24	1.4	1.2	13	18	17
Western Europe	0.15	0.16	0.16	177	201	187	1.1	1.1	38	41	40
OECD	**0.12**	**0.13**	**0.13**	**286**	**349**	**344**	**1.2**	**1.2**	**31**	**34**	**34**
World	**0.08**	**0.11**	**0.13**	**1,786**	**3,655**	**4,682**	**2.0**	**2.6**	**32**	**47**	**53**
Policy Reform											
Africa	0.04	0.06	0.08	144	346	493	2.4	3.4	21	27	29
China+	0.16	0.15	0.14	284	281	243	1.0	0.9	21	17	15
Latin America	0.02	0.02	0.02	41	65	82	1.6	2.0	9	10	11
Middle East	0.62	0.59	0.58	154	284	362	1.8	2.4	89	88	88
South Asia	0.26	0.31	0.35	688	1,194	1,519	1.7	2.2	56	66	74
Southeast Asia	0.02	0.03	0.03	52	103	123	2.0	2.4	11	17	18
Developing	**0.07**	**0.08**	**0.08**	**1,363**	**2,273**	**2,822**	**1.7**	**2.1**	**31**	**36**	**39**
Eastern Europe	0.31	0.24	0.21	58	42	32	0.7	0.6	59	46	41
FSU	0.07	0.05	0.04	79	67	56	0.8	0.7	27	23	21
Transitional	**0.08**	**0.06**	**0.05**	**137**	**109**	**88**	**0.8**	**0.6**	**35**	**29**	**25**
North America	0.11	0.08	0.06	89	63	28	0.7	0.3	30	17	7
Pac OECD	0.08	0.07	0.06	20	15	10	0.8	0.5	13	10	7
Western Europe	0.15	0.09	0.07	177	120	84	0.7	0.5	38	24	18
OECD	**0.12**	**0.08**	**0.07**	**286**	**198**	**122**	**0.7**	**0.4**	**31**	**20**	**12**
World	**0.08**	**0.08**	**0.08**	**1,786**	**2,580**	**3,032**	**1.4**	**1.7**	**32**	**34**	**36**

P-4 Carbon emissions

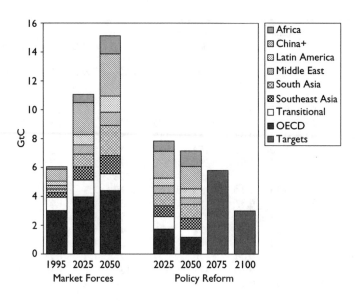

P-4 Carbon emissions

	Total annual emissions (GtC)			Index (1995 = 1)		Annual per capita (tC)			Annual per dollar GDP (kgC)		
	1995	2025	2050	2025	2050	1995	2025	2050	1995	2025	2050
Market Forces											
Africa	0.17	0.57	1.25	3.4	7.4	0.2	0.4	0.7	0.12	0.13	0.13
China+	0.83	2.21	2.93	2.7	3.5	0.6	1.3	1.8	0.21	0.18	0.14
Latin America	0.30	0.71	1.12	2.4	3.7	0.6	1.0	1.4	0.11	0.10	0.08
Middle East	0.25	0.64	0.91	2.6	3.6	1.4	2.0	2.1	0.20	0.16	0.11
South Asia	0.25	0.90	2.08	3.6	8.3	0.2	0.5	1.0	0.13	0.14	0.14
Southeast Asia	0.32	0.91	1.27	2.8	4.0	0.7	1.4	1.8	0.13	0.12	0.09
Developing	**2.12**	**5.94**	**9.56**	**2.8**	**4.5**	**0.5**	**0.9**	**1.3**	**0.15**	**0.14**	**0.12**
Eastern Europe	0.19	0.24	0.22	1.3	1.2	1.9	2.6	2.7	0.34	0.25	0.19
FSU	0.73	0.92	0.94	1.3	1.3	2.5	3.1	3.3	0.67	0.46	0.35
Transitional	**0.92**	**1.16**	**1.16**	**1.3**	**1.3**	**2.4**	**3.0**	**3.2**	**0.56**	**0.40**	**0.30**
North America	1.60	2.24	2.60	1.4	1.6	5.3	6.1	6.5	0.20	0.12	0.09
Pac OECD	0.40	0.48	0.48	1.2	1.2	2.7	3.1	3.4	0.13	0.08	0.06
Western Europe	1.01	1.24	1.32	1.2	1.3	2.2	2.5	2.8	0.12	0.08	0.06
OECD	**3.01**	**3.96**	**4.40**	**1.3**	**1.5**	**3.3**	**3.9**	**4.4**	**0.16**	**0.10**	**0.07**
World	**6.05**	**11.06**	**15.12**	**1.8**	**2.5**	**1.1**	**1.4**	**1.7**	**0.17**	**0.13**	**0.10**
Policy Reform											
Africa	0.17	0.71	1.09	4.2	6.4	0.2	0.6	0.6	0.12	0.11	0.07
China+	0.83	1.87	1.55	2.3	1.9	0.6	1.2	1.0	0.21	0.14	0.07
Latin America	0.30	0.53	0.62	1.8	2.1	0.6	0.8	0.8	0.11	0.07	0.04
Middle East	0.25	0.52	0.45	2.1	1.8	1.4	1.6	1.1	0.20	0.13	0.06
South Asia	0.25	0.86	0.95	3.4	3.8	0.2	0.5	0.5	0.13	0.11	0.05
Southeast Asia	0.32	0.75	0.76	2.3	2.4	0.7	1.2	1.1	0.13	0.09	0.04
Developing	**2.12**	**5.24**	**5.42**	**2.5**	**2.6**	**0.5**	**0.8**	**0.8**	**0.15**	**0.11**	**0.06**
Eastern Europe	0.19	0.17	0.11	0.9	0.6	1.9	1.8	1.4	0.34	0.16	0.08
FSU	0.73	0.69	0.47	0.9	0.6	2.5	2.4	1.7	0.67	0.29	0.14
Transitional	**0.92**	**0.86**	**0.58**	**0.9**	**0.6**	**2.4**	**2.3**	**1.7**	**0.56**	**0.25**	**0.12**
North America	1.60	0.92	0.60	0.6	0.4	5.3	2.5	1.5	0.20	0.07	0.04
Pac OECD	0.40	0.22	0.14	0.6	0.4	2.7	1.4	1.0	0.13	0.05	0.03
Western Europe	1.01	0.60	0.42	0.6	0.4	2.2	1.2	0.9	0.12	0.04	0.03
OECD	**3.01**	**1.74**	**1.16**	**0.6**	**0.4**	**3.3**	**1.7**	**1.2**	**0.16**	**0.06**	**0.03**
World	**6.05**	**7.84**	**7.16**	**1.3**	**1.2**	**1.1**	**1.0**	**0.8**	**0.17**	**0.09**	**0.05**

P-5 Sulfur emissions

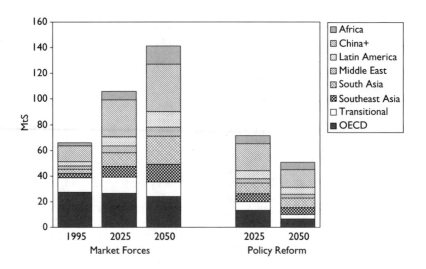

P-5 Sulfur emissions

	Emissions (MtS)			Growth rate (%/year)			Index (1995 = 1)	
	1995	2025	2050	1995–2025	2025–2050	1995–2050	2025	2050
Market Forces								
Africa	2	6	14	3.7	3.1	3.4	3.0	6.4
China+	13	29	37	2.8	1.0	2.0	2.3	3.0
Latin America	3	7	12	2.7	2.1	2.4	2.2	3.7
Middle East	3	5	7	2.5	1.1	1.9	2.1	2.8
South Asia	4	11	22	3.7	2.9	3.3	3.0	6.0
Southeast Asia	3	8	14	3.5	1.9	2.8	2.8	4.6
Developing	**27**	**67**	**106**	**3.1**	**1.8**	**2.5**	**2.5**	**3.9**
Eastern Europe	2	3	2	0.7	−0.3	0.2	1.2	1.1
FSU	9	10	9	0.2	−0.3	0.0	1.1	1.0
Transitional	**11**	**12**	**11**	**0.3**	**−0.3**	**0.0**	**1.1**	**1.0**
North America	13	13	12	0.0	−0.2	−0.1	1.0	1.0
Pac OECD	5	5	4	−0.3	−0.8	−0.5	0.9	0.8
Western Europe	10	9	8	−0.2	−0.5	−0.3	0.9	0.8
OECD	**28**	**27**	**24**	**−0.1**	**−0.4**	**−0.2**	**1.0**	**0.9**
World	**66**	**106**	**141**	**1.6**	**1.2**	**1.4**	**1.6**	**2.1**
Policy Reform								
Africa	2	6	6	3.5	0.0	1.9	2.8	2.8
China+	13	21	14	1.8	−1.8	0.1	1.7	1.1
Latin America	3	6	6	2.2	−0.4	1.0	1.9	1.7
Middle East	3	4	3	1.0	−0.6	0.3	1.4	1.2
South Asia	4	8	7	2.6	−0.3	1.3	2.2	2.0
Southeast Asia	3	6	6	2.6	−0.5	1.2	2.1	1.9
Developing	**27**	**51**	**41**	**2.1**	**−0.9**	**0.8**	**1.9**	**1.5**
Eastern Europe	2	1	1	−1.6	−2.8	−2.2	0.6	0.3
FSU	9	5	2	−1.8	−3.1	−2.4	0.6	0.3
Transitional	**11**	**7**	**3**	**−1.7**	**−3.0**	**−2.3**	**0.6**	**0.3**
North America	13	6	3	−2.5	−2.6	−2.6	0.5	0.2
Pac OECD	5	3	1	−2.0	−3.1	−2.5	0.5	0.3
Western Europe	10	5	2	−2.4	−2.6	−2.5	0.5	0.2
OECD	**28**	**13**	**7**	**−2.4**	**−2.7**	**−2.5**	**0.5**	**0.2**
World	**66**	**71**	**51**	**0.3**	**−1.3**	**−0.5**	**1.1**	**0.8**

P-6 Forestry

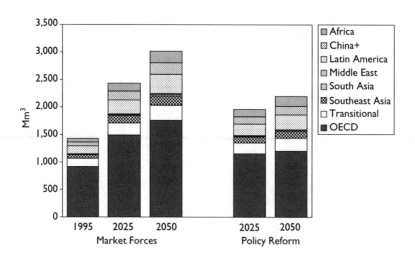

P-6 Forestry

	Forest resources										
	Production from forests (Mm³)			1995 Area (Mha)		Exploitation intensity (%)			Plantation (Mha)		
	1995	2025	2050	Total	Exploitable	1995	2025	2050	1995	2025	2050
Market Forces											
Africa	66	144	208	709	166	55	90	95	4	11	19
China+	71	166	210	154	74	35	55	45	20	26	26
Latin America	143	250	350	1,000	188	64	90	95	6	10	16
Middle East	7	13	18	18	8	110	90	95	0	9	8
South Asia	8	20	29	77	26	50	90	95	4	7	12
Southeast Asia	66	133	168	255	126	64	90	95	5	16	25
Developing	361	725	982	2,213	589	55	81	82	39	79	107
Eastern Europe	49	67	84	25	22	61	82	95	1	1	4
FSU	106	152	191	788	533	6	6	6	22	23	29
Transitional	154	218	275	813	555	9	10	11	23	25	33
North America	592	963	1,143	731	304	75	90	95	18	61	74
Pac OECD	61	153	181	170	24	8	8	8	13	17	21
Western Europe	259	374	431	142	99	61	88	95	9	12	17
OECD	912	1,490	1,756	1,043	426	68	87	93	41	91	112
World	1,427	2,433	3,013	4,069	1,570	37	46	48	103	194	251
Policy Reform											
Africa	66	135	181	709	166	55	70	70	4	12	18
China+	71	132	154	154	74	35	38	27	20	22	20
Latin America	143	211	273	1,000	188	64	70	70	6	9	13
Middle East	7	7	4	18	8	110	57	59	0	4	0
South Asia	8	16	21	77	26	50	70	70	4	6	8
Southeast Asia	66	108	127	255	126	64	70	70	5	13	18
Developing	361	609	759	2,213	589	55	61	58	39	65	76
Eastern Europe	49	60	72	25	22	61	70	70	1	2	5
FSU	106	136	164	788	533	6	6	6	22	22	22
Transitional	154	197	235	813	555	9	9	9	23	25	27
North America	592	722	756	731	304	75	70	70	18	39	38
Pac OECD	61	122	128	170	24	8	8	8	13	14	14
Western Europe	259	310	317	142	99	61	70	70	9	11	10
OECD	912	1,154	1,202	1,043	426	68	68	68	41	63	63
World	1,427	1,959	2,196	4,069	1,570	37	37	37	103	154	166

P-7 Land use patterns

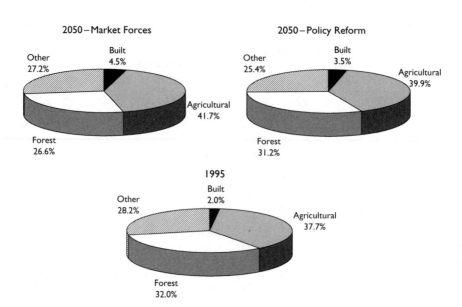

2050 – Market Forces

Other
27.2%

Built
4.5%

Agricultural
41.7%

Forest
26.6%

2050 – Policy Reform

Other
25.4%

Built
3.5%

Agricultural
39.9%

Forest
31.2%

1995

Other
28.2%

Built
2.0%

Agricultural
37.7%

Forest
32.0%

P-7 Land use patterns

| | Total land area (Mha) | Share of land area (%) | | | | | | | | | | | |
| | | 1995 | | | | 2025 | | | | 2050 | | | |
		Built	Agri-culture	Forest	Other	Built	Agri-culture	Forest	Other	Built	Agri-culture	Forest	Other
Market Forces													
Africa	2,937	2	37	24	38	3	39	20	38	4	42	16	38
China+	1,157	3	57	15	24	6	63	10	21	8	61	10	22
Latin America	2,016	1	37	50	11	2	47	40	10	4	53	33	10
Middle East	603	2	45	3	51	3	44	3	50	5	43	3	49
South Asia	408	10	55	20	15	16	51	17	16	23	45	17	15
Southeast Asia	440	3	23	59	15	7	23	56	14	12	22	54	13
Developing	**7,561**	**2**	**41**	**30**	**27**	**4**	**45**	**24**	**26**	**6**	**47**	**21**	**26**
Eastern Europe	89	4	60	29	6	5	54	30	11	5	52	30	13
FSU	2,195	1	27	37	36	1	28	36	35	1	30	36	34
Transitional	**2,284**	**1**	**28**	**37**	**35**	**1**	**29**	**36**	**34**	**1**	**31**	**36**	**33**
North America	1,840	2	27	41	30	3	27	41	30	3	27	40	30
Pac OECD	841	1	58	22	20	1	57	22	20	1	57	22	20
Western Europe	495	6	41	31	23	7	41	30	22	7	43	30	21
OECD	**3,176**	**2**	**37**	**34**	**26**	**3**	**37**	**34**	**26**	**3**	**37**	**34**	**26**
World	**13,021**	**2**	**38**	**32**	**28**	**3**	**40**	**29**	**28**	**4**	**42**	**27**	**27**
Policy Reform													
Africa	2,937	2	37	24	38	3	38	22	37	3	38	22	36
China+	1,157	3	57	15	24	5	62	14	19	7	61	14	19
Latin America	2,016	1	37	50	11	2	44	46	8	2	45	47	6
Middle East	603	2	45	3	51	3	44	3	50	4	43	3	50
South Asia	408	10	55	20	15	16	51	18	15	21	47	19	13
Southeast Asia	440	3	23	59	15	6	22	59	12	9	21	61	9
Developing	**7,561**	**2**	**41**	**30**	**27**	**4**	**43**	**28**	**25**	**5**	**43**	**28**	**24**
Eastern Europe	89	4	60	29	6	5	53	31	12	5	50	33	12
FSU	2,195	1	27	37	36	1	28	37	35	1	30	37	33
Transitional	**2,284**	**1**	**28**	**37**	**35**	**1**	**29**	**36**	**34**	**1**	**30**	**36**	**32**
North America	1,840	2	27	41	30	2	28	41	28	2	29	42	27
Pac OECD	841	1	58	22	20	1	57	22	20	1	56	22	21
Western Europe	495	6	41	31	23	6	42	31	22	5	45	31	19
OECD	**3,176**	**2**	**37**	**34**	**26**	**2**	**38**	**35**	**25**	**2**	**38**	**35**	**24**
World	**13,021**	**2**	**38**	**32**	**28**	**3**	**40**	**31**	**27**	**4**	**40**	**31**	**25**

P-8 Land use change

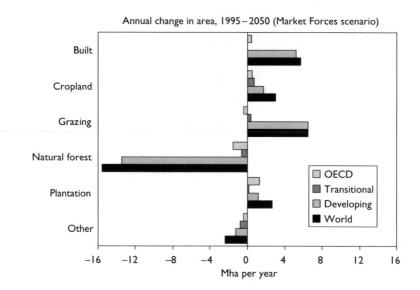

Annual change in area, 1995 – 2050 (Market Forces scenario)

P-8 Land use change

Region	Land use	1995		2050		Annual change 1995–2050 (Mha/year)
		Area (Mha)	Share (%)	Area (Mha)	Share (%)	
OECD		**3,176**	**100**	**3,176**	**100**	**0.0**
	Built environment	76	2	103	3	0.5
	Cropland	406	13	434	14	0.5
	Grazing	776	24	755	24	−0.4
	Natural forest	1,043	33	959	30	−1.5
	Plantation	41	1	112	4	1.3
	Protected (non-forest)	283	9	283	9	0.0
	All other	551	17	530	17	−0.4
Transitional		**2,284**	**100**	**2,284**	**100**	**0.0**
	Built environment	16	1	19	1	0.1
	Cropland	268	12	307	13	0.7
	Grazing	370	16	390	17	0.4
	Natural forest	813	36	781	34	−0.6
	Plantation	23	1	33	1	0.2
	Protected (non-forest)	8	0	8	0	0.0
	All Other	786	34	745	33	−0.7
Developing		**7,561**	**100**	**7,561**	**100**	**0.0**
	Built environment	175	2	460	6	5.2
	Cropland	829	11	925	12	1.8
	Grazing	2,261	30	2,620	35	6.5
	Natural forest	2,213	29	1,474	19	−13.4
	Plantation	39	1	105	1	1.2
	Protected (non-forest)	278	4	278	4	0.0
	All other	1,766	23	1,698	22	−1.2
World		**13,021**	**100**	**13,021**	**100**	**0.0**
	Built environment	267	2	583	4	5.7
	Cropland	1,503	12	1,667	13	3.0
	Grazing	3,406	26	3,764	29	6.5
	Natural forest	4,069	31	3,213	25	−15.6
	Plantation	103	1	250	2	2.7
	Protected (non-forest)	570	4	570	4	0.0
	All other	3,102	24	2,973	23	−2.3

P-8 Land use change (*continued*)

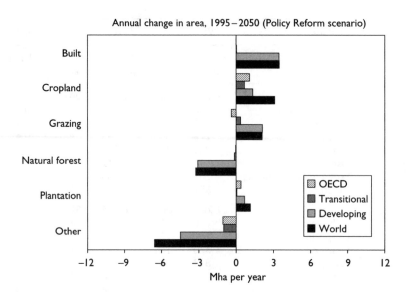

Annual change in area, 1995–2050 (Policy Reform scenario)

P-8 Land use change (*continued*)

Region	Land use	1995		2050		Annual change 1995–2050
		Area (Mha)	Share (%)	Area (Mha)	Share (%)	(Mha/yr)
OECD		**3,176**	**100**	**3,176**	**100**	**0.0**
	Built environment	76	2	76	2	0.0
	Cropland	406	13	466	15	1.1
	Grazing	776	24	755	24	−0.4
	Natural forest	1,043	33	1,042	33	0.0
	Plantation	41	1	63	2	0.4
	Protected (non-forest)	283	9	283	9	0.0
	All other	551	17	492	15	−1.1
Transitional		**2,284**	**100**	**2,284**	**100**	**0.0**
	Built environment	16	1	18	1	0.0
	Cropland	268	12	305	13	0.7
	Grazing	370	16	390	17	0.4
	Natural forest	813	36	806	35	−0.1
	Plantation	23	1	27	1	0.1
	Protected (non-forest)	8	0	8	0	0.0
	All Other	786	34	731	32	−1.0
Developing		**7,561**	**100**	**7,561**	**100**	**0.0**
	Built environment	175	2	365	5	3.4
	Cropland	829	11	903	12	1.3
	Grazing	2,261	30	2,376	31	2.1
	Natural forest	2,213	29	2,044	27	−3.1
	Plantation	39	1	76	1	0.7
	Protected (non-forest)	278	4	278	4	0.0
	All Other	1,766	23	1,519	20	−4.5
World		**13,021**	**100**	**13,021**	**100**	**0.0**
	Built environment	267	2	458	4	3.5
	Cropland	1,503	12	1,673	13	3.1
	Grazing	3,406	26	3,521	27	2.1
	Natural forest	4,069	31	3,891	30	−3.2
	Plantation	103	1	166	1	1.1
	Protected (non-forest)	570	4	570	4	0.0
	All other	3,102	24	2,742	21	−6.6

P-9 Nitrogen fertilizer

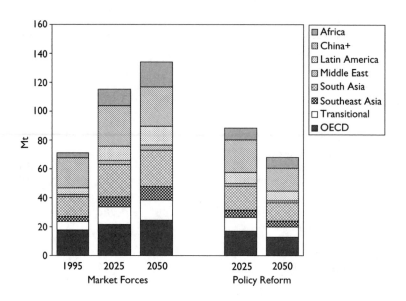

P-9 Nitrogen fertilizer

	Consumption (Mt)			Growth rate (%/year)			Index (1995 = 1)	
	1995	2025	2050	1995– 2025	2025– 2050	1995– 2050	2025	2050
Market Forces								
Africa	4	12	17	4.0	1.6	2.9	3.3	4.9
China+	21	28	27	1.0	−0.1	0.5	1.3	1.3
Latin America	5	10	13	2.5	1.1	1.9	2.1	2.8
Middle East	1	3	4	2.3	0.9	1.7	2.0	2.5
South Asia	14	22	25	1.7	0.5	1.1	1.6	1.8
Southeast Asia	3	7	9	2.4	1.2	1.8	2.0	2.7
Developing	**48**	**82**	**96**	**1.8**	**0.6**	**1.3**	**1.7**	**2.0**
Eastern Europe	1	2	2	0.9	0.5	0.7	1.3	1.5
FSU	5	11	12	2.7	0.6	1.7	2.2	2.5
Transitional	**6**	**12**	**14**	**2.4**	**0.6**	**1.5**	**2.0**	**2.3**
North America	8	10	12	0.7	0.5	0.6	1.2	1.4
Pac OECD	1	2	2	1.3	0.7	1.0	1.5	1.7
Western Europe	8	9	11	0.5	0.5	0.5	1.2	1.3
OECD	**18**	**22**	**24**	**0.7**	**0.5**	**0.6**	**1.2**	**1.4**
World	**71**	**115**	**134**	**1.6**	**0.6**	**1.2**	**1.6**	**1.9**
Policy Reform								
Africa	4	8	8	2.9	−0.3	1.4	2.3	2.1
China+	21	23	16	0.3	−1.4	−0.5	1.1	0.8
Latin America	5	8	7	1.7	−0.6	0.6	1.7	1.4
Middle East	1	2	1	0.8	−0.9	0.0	1.3	1.0
South Asia	14	17	13	0.6	−1.1	−0.1	1.2	0.9
Southeast Asia	3	5	4	1.3	−0.9	0.3	1.5	1.2
Developing	**48**	**62**	**48**	**0.9**	**−1.0**	**0.0**	**1.3**	**1.0**
Eastern Europe	1	1	1	0.1	−1.5	−0.6	1.0	0.7
FSU	5	8	6	1.9	−1.1	0.5	1.7	1.3
Transitional	**6**	**10**	**7**	**1.6**	**−1.2**	**0.3**	**1.6**	**1.2**
North America	8	8	6	−0.1	−1.1	−0.5	1.0	0.7
Pac OECD	1	1	1	0.3	−1.3	−0.4	1.1	0.8
Western Europe	8	7	6	−0.3	−1.1	−0.7	0.9	0.7
OECD	**18**	**17**	**13**	**−0.1**	**−1.1**	**−0.6**	**1.0**	**0.7**
World	**71**	**88**	**68**	**0.7**	**−1.0**	**−0.1**	**1.2**	**1.0**

P-10 Fisheries

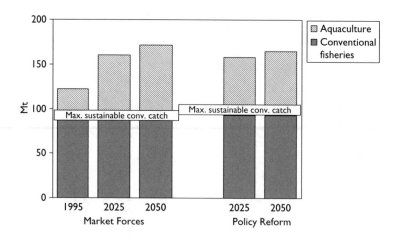

P-10 Fisheries

	Production (Mt)			Growth rate (%/year)			Index (1995 = 1)	
	1995	2025	2050	1995– 2025	2025– 2050	1995– 2050	2025	2050
Market Forces								
Conventional fisheries	88	88	88	0.0	0.0	0.0	1.0	1.0
Aquaculture	34	72	84	2.5	0.6	1.6	2.1	2.4
Total	**122**	**160**	**172**	**0.9**	**0.3**	**0.6**	**1.3**	**1.4**
Policy Reform								
Conventional fisheries	88	93	93	0.2	0.0	0.1	1.1	1.1
Aquaculture	34	65	72	2.1	0.4	1.4	1.9	2.1
Total	**122**	**158**	**165**	**0.9**	**0.2**	**0.5**	**1.3**	**1.3**

P-11 Toxic waste

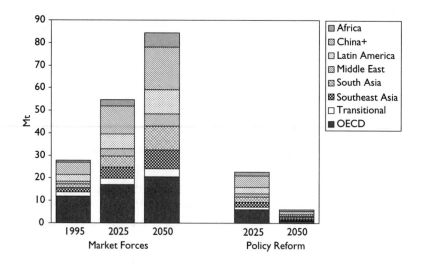

P-11 Toxic waste

	Toxic waste (Mt)			Growth rate (%/year)			Index (1995 = 1)	
	1995	2025	2050	1995–2025	2025–2050	1995–2050	2025	2050
Market Forces								
Africa	0.8	2.7	6.3	4.1	3.4	3.8	3.3	7.7
China+	5.4	12.4	19.0	2.8	1.7	2.3	2.3	3.5
Latin America	3.0	6.5	10.8	2.6	2.0	2.3	2.2	3.5
Middle East	1.2	3.2	5.4	3.2	2.1	2.7	2.6	4.3
South Asia	1.6	4.9	10.5	3.7	3.1	3.4	3.0	6.4
Southeast Asia	1.8	4.9	8.4	3.4	2.2	2.8	2.7	4.6
Developing	**14**	**35**	**60**	**3.1**	**2.2**	**2.7**	**2.5**	**4.3**
Eastern Europe	0.5	0.8	0.9	1.5	0.7	1.1	1.6	1.9
FSU	1.4	2.0	2.5	1.2	0.9	1.1	1.4	1.8
Transitional	**2**	**3**	**3**	**1.3**	**0.8**	**1.1**	**1.5**	**1.8**
North America	4.0	6.1	7.7	1.4	0.9	1.2	1.5	1.9
Pac OECD	2.3	3.1	3.4	0.9	0.4	0.7	1.3	1.5
Western Europe	5.6	8.0	9.6	1.2	0.7	1.0	1.4	1.7
OECD	**12**	**17**	**21**	**1.2**	**0.7**	**1.0**	**1.4**	**1.7**
World	**28**	**55**	**84**	**2.3**	**1.8**	**2.0**	**2.0**	**3.0**
Policy Reform								
Africa	0.8	1.7	0.7	2.4	−3.4	−0.3	2.0	0.9
China+	5.4	5.3	1.4	−0.1	−5.3	−2.5	1.0	0.3
Latin America	3.0	2.8	0.8	−0.3	−4.9	−2.4	0.9	0.3
Middle East	1.2	1.3	0.4	0.1	−4.8	−2.2	1.0	0.3
South Asia	1.6	2.3	0.9	1.2	−3.9	−1.2	1.4	0.5
Southeast Asia	1.8	2.1	0.6	0.6	−4.8	−1.9	1.2	0.3
Developing	**14**	**15**	**5**	**0.3**	**−4.6**	**−2.0**	**1.1**	**0.3**
Eastern Europe	0.5	0.3	0.1	−1.3	−6.1	−3.5	0.7	0.1
FSU	1.4	0.9	0.2	−1.3	−5.7	−3.4	0.7	0.2
Transitional	**2**	**1**	**0**	**−1.3**	**−5.8**	**−3.4**	**0.7**	**0.1**
North America	4.0	2.1	0.4	−2.2	−6.4	−4.1	0.5	0.1
Pac OECD	2.3	1.1	0.2	−2.6	−6.9	−4.5	0.5	0.1
Western Europe	5.6	2.9	0.5	−2.2	−6.6	−4.2	0.5	0.1
OECD	**12**	**6**	**1**	**−2.3**	**−6.6**	**−4.2**	**0.5**	**0.1**
World	**28**	**23**	**6**	**−0.7**	**−5.1**	**−2.7**	**0.8**	**0.2**

Abbreviations

GDP	gross domestic product
MER	market exchange rate
PPP	purchasing power parity
GDP$_{MER}$	GDP in US dollars at MER
GDP$_{PPP}$	GDP in international dollars adjusted for PPP
OECD	Organization for Economic Cooperation and Development
FSU	Former Soviet Union
PCL	potential cultivable land

°C	degree centigrade
bbl	barrel
cal	calorie
g	gram
ha	hectare
Mha	million hectares
m	meter
mpg	miles per gallon
t	tonne
Mt	million tonnes
tC	tonne of carbon
J	Joule
BOE	barrel of oil equivalent
TOE	tonne of oil equivalent
TCE	tonne of coal equivalent

E	Exa, 10^{18}
P	Peta, 10^{15}
T	Tera, 10^{12}
G	Giga, 10^9
M	Mega, 10^6
k	kilo, 10^3

Selected conversion factors

Energy: $1\,GJ = 0.172\,BOE = 0.0239\,TOE = 0.0341\,TCE$
Volume: $1\,km^3 = 264 \times 10^9\,US\ gallons = 811 \times 10^3\,acre\ feet$
Mass: $1\,t = 1,000\,kg = 1.10\,short\ t = 2.20 \times 10^3\,pounds$
Area: $1\,ha = 2.47\,acre = 0.01\,km^2$
Vehicle efficiency: $1\,km/l = 2.35\,mpg\ (US)$

Annex notes

Demography

D-1: Base year data from UN (1999). In the OECD regions in both scenarios, populations are given by the mid-range UN projections (1999). In non-OECD regions, *Market Forces* populations are from the mid-range UN projections, while in the *Policy Reform* scenario, higher incomes and targeted policies lead to populations 3 percent below UN mid-range in 2025 and 5 percent below the mid-range projections in 2050 (see Chapters 4 and 5).

D-2: Values for 1995 and 2025 based on WRI (1994, 1996b), with continued urbanization thereafter at a moderated rate in both scenarios.

Economy

E-1: 1995 GDP_{MER} and GDP_{PPP} from WRI (1998). The *Market Forces* scenario follows typical mid-range patterns (World Bank 1991, 1999, IPCC 1992, OECD 2001) (see Chapter 4). In the *Policy Reform* scenario, changes in income and income distribution (Annex S-1) are constrained by the sustainability goal of reducing poverty (see Chapter 5).

E-2: 1995 sectoral GDP shares based on WRI (1998). The scenarios for GDP structure reflect global convergence toward the pattern of declining agricultural share, initially rising and then slowly declining industrial share, and rising services share, seen historically in industrialized countries.

E-3: Income is represented by GDP per capita. Figures in the table are rounded to the nearest $100. For 1995 figures, regional incomes are population-weighted averages of country incomes (WRI 1998). The *Market Forces* scenario follows typical mid-range patterns (World Bank 1991, IPCC 1992) (see Chapter 4). *Market Forces* scenario growth rates for GDP_{PPP} are based on GDP_{MER} growth rates (IPCC 1992) (see Chapter 4). In the *Policy Reform* scenario, changes in income and income distribution (Annex S-1) are constrained by the sustainability goal of reducing poverty (see Chapter 5 for details). In the OECD regions, differences in values and lifestyle choices between the *Market Forces* and *Policy Reform* scenarios lead to relatively lower income growth in *Policy Reform*. In non-OECD countries, equity considerations, and development patterns characterized by convergence to industrialized country norms, are reflected in faster income growth in *Policy Reform* relative to *Market Forces*.

Society

S-1: Regional values are computed from national values weighted by population. Gini coefficients (see Chapter 4 for definition) are from Deininger and Squire (1996), Tabatabai (1996), USBC (1997), UNU/WIDER (1999), and World Bank (2000). Ratios of the average income of the low-est-earning 20 percent to the highest-earning 20 percent of the population ("national equity") are estimated from Gini coefficients assuming lognormal national income distributions (see

Kemp-Benedict *et al.* 2002). In *Market Forces*, income distributions become more unequal in nearly every region, as countries converge to the evolving US pattern (see Chapter 4). In the *Policy Reform* scenario, changes in income (Annex E-3) and income distribution are constrained by the sustainability goal of reducing poverty (see Chapter 5). Income distributions in the scenario converge toward a common value of greater equality than in the *Market Forces* scenario. This pattern leads to rising inequality in countries and regions where it is currently low, and falling inequality where it is currently high, but in all cases equity is greater in *Policy Reform* than in *Market Forces*.

S-2: Developing and transitional regions, 1995 data based on FAO estimates of the national incidence of chronic undernutrition (1999c, 2000e); for the US, the incidence of food insecurity was used (Rose *et al.* 1995); for all other countries estimates are based on income distribution (Kemp-Benedict *et al.* 2002). In scenarios, hunger patterns are determined by changes in income (Annex E-3), income distribution (Annex S-1) and population (Annex D-1) (see Chapter 4). The *Policy Reform* scenario meets the sustainability goal of reducing hunger by one-half between 1995 and 2025 and by one-half again between 2025 and 2050 (for sustainability goals see Chapter 3). The goal is met through a combination of measures to moderate population growth, increase income growth and decrease inequality relative to *Market Forces* (see Chapter 5).

Energy

En-1: Data are from IEA energy statistics (IEA 1999). Primary energy in each region is the sum of final energy demand in the region, energy lost through the transformation of primary to secondary fuel forms, and net energy lost or used during transmission and distribution. Care should be taken in interpreting primary energy supply figures since different statistical sources use different conventions for reporting the primary energy content of nuclear and renewable sources of energy. The conventions for efficiency adopted in this study are: uranium at 33 percent; geothermal at 10 percent; and hydro, wind, solar, and other renewables at 100 percent. Other sources may express hydro and other resources in terms of an equivalent fossil fuel feedstock requirement.

In the scenarios, primary energy use is driven by changing final energy demands (Annexes En-3 and En-4), energy conversion efficiencies and fuel mix used in electricity generation. In the OECD regions, final energy demand (Annex En-3) is lower in *Policy Reform* than in *Market Forces*, as discussed in the notes for Annex En-3. Conversion efficiencies increase in all regions in both scenarios, with more rapid increases in the *Policy Reform* scenario. Electricity generation fuel mixes shift toward less carbon-intensive fuels in both scenarios, toward natural gas, and away from coal and oil. In *Policy Reform*, these changes are accompanied by increasing use of renewables in electricity generation. Nuclear electricity generation continues to play a role in the *Market Forces* scenario, but in *Policy Reform* it is phased out by 2050 (see Chapter 5).

En-2: Both scenarios have an increased share of natural gas. In the *Market Forces* scenario, declines in traditional biomass use, with minimal investment in modern renewables (which includes biomass, as well as geothermal, solar, and wind), lead to a gradual absolute increase in renewables, but a steady drop in share of total output (see Chapter 4). Along with efficiency increases (Annex En-1), the sustainability goal of limiting climate change is met in large part in the *Policy Reform* scenario through switching from coal and oil to a mix of natural gas and renewables, while uranium use for electricity production is steadily reduced, falling to zero by 2050 (see Chapter 5).

En-3: Data are from IEA energy statistics (IEA 1999). Final energy demand is defined as consumption of electricity and fuels by end users. In the OECD regions, final energy demand is lower in *Policy Reform* than in *Market Forces*, due in part to lower GDP and in part to lower energy intensity, the amount of energy consumed to produce one dollar of end-use service. Note that the energy intensity reported in the table is an aggregate measure (MJ/$PPP) that is affected by energy efficiency, the composition of economic activity and energy-use patterns. In non-OECD regions, rapid economic increases drive energy demand growth, which is partially offset by improved efficiency. In *Policy Reform*, higher economic growth rates are offset by high efficiencies as developing regions more rapidly converge to OECD standards.

En-4: Several factors influence sector shares. In all sectors, energy intensity decreases, to a greater degree in *Policy Reform* than in *Market Forces*. Agricultural output declines as a share in total economic output, with a corresponding decline in its share of energy demand. Household energy use rises rapidly in developing regions as households modernize. The composition of industrial output continues to shift to one less dominated by heavy, energy-intensive subsectors. In transport, the increasing dominance of private vehicles in *Market Forces* is moderated in *Policy Reform*.

En-5: 1995 data are from the WEC Survey of Energy Resources (WEC 1995). *Proven* reserves are those that have been demonstrated to exist. *Undiscovered* reserves are reserves that are thought likely to exist and to be potentially exploitable, based on existing geological and economic information. Reserves are reported here in physical terms – tonnes of coal and uranium, barrels (bbl) of oil and cubic meters of natural gas. (See page 198 for conversion factors.) Conversion factors used to estimate uranium reserves assume conventional fission reactors. Estimates for petroleum and natural gas do not include nonconventional oil and natural gas resources (see Chapter 4). Thus, it is possible for consumption to exceed available conventional reserves, as in the *Market Forces* scenario.

Food and agriculture

F-1: Average daily consumption includes levels of human intake, food waste and other losses. Since lifestyles do not vary between the two *Conventional Worlds* scenarios, OECD consumption is the same in both scenarios. North America and Western Europe remain at their relatively high 1995 values, while in Pacific OECD, caloric intake rises from its low 1995 level. In North America, meat consumption as a fraction of total calories declines slightly due to health concerns. In the non-OECD regions, consumption rises with income. Higher income levels in *Policy Reform* lead to slightly higher food consumption relative to *Market Forces*. The share of calories derived from animal products also rises in non-OECD regions, with a higher increase in *Policy Reform* compared to *Market Forces* (see Chapter 5).

F-2: Requirements, production and trade based on FAO (1999b). The self-sufficiency ratio, a measure of import dependence, is calculated as production divided by requirements. Net exports in the base year include net additions to stocks. Meat includes eggs and meat byproducts, and milk includes all products derived from milk. Meat and milk consumption is higher in the *Policy Reform* scenario than in the *Market Forces* scenario, despite lower populations in *Policy Reform*, driven by higher incomes in developing regions. Feedlot production is an indicator of industrialization in livestock production. It is computed as the proportion of total animal feed supplied by crop and animal product concentrates (feed grain, oilcakes, fishmeal, offals, etc.). The rest comes from grazing, fodder, crop residues, and household wastes. 1995 values are estimated from crop production and animal stocks data (FAO 1999b), feed energy contents (Lalonde and Sukigara 1997, FAO 2000g) and estimates of animal feed requirements (IPCC/OECD/IEA 1996, Lalonde and Sukigara 1997) (see Kemp-Benedict *et al.* 2002). Note that any particular animal might be fed from a variety of sources over its lifetime. Although grazing efficiencies improve in the scenarios (especially in *Policy Reform*), feedlot production rises sharply in most regions in both scenarios, due to rapidly growing demands and limits on the expansion of grazing land. Feedlot production is higher in *Policy Reform* than in *Market Forces*, driven by higher demand and tighter restrictions on the expansion of agricultural land.

F-3: 1995 figures based on FAO (1999b). "Fish" includes all seafood. The self-sufficiency ratio, a measure of import dependence, is calculated as production divided by requirements. Pressure on natural fish stocks limits increases in fish consumption and production (see Annex P-10). Per-capita consumption is similar in the two scenarios, with lower consumption in *Policy Reform* due mainly to lower populations.

F-4: 1995 figures based on FAO (1999b). The self-sufficiency ratio, a measure of import dependence, is calculated as production divided by requirements. Net exports in the base year include net additions to stocks. Crop requirements rise in both scenarios, driven by increases in food and feed demand, with both higher in *Policy Reform* (see Annexes F-1 and F-2). In both scenarios, self-sufficiency ratios tend to remain close to base-year levels for importers, as production rises in proportion to demand, unless land constraints limit expanded production. Constraints are more limiting under *Policy Reform*, where some land is set aside for ecosystem protection. Exporters expand production to meet increased import requirements, based mainly on 1995 trade patterns. Production in the transitional regions increases considerably, as yields recover from the sharp drop in the early 1990s (Annex F-5).

F-5: 1995 figures based on FAO (1999b). Annual yield is defined as total annual production per unit area of cropland. Harvest yield is production per unit area per harvest. Cropping intensity, a measure of multiple cropping, is the average number of harvests per year (annual yield = cropping intensity × harvest yield). Cropping intensity is greater than one if more than one crop is planted per year, and less than one if land is left fallow for some years. Yields and cropping intensity values are averages over irrigated and rainfed practices. In general, yields are higher on irrigated land than on rainfed land, so shifts in the composition of cropland affect average yields. In the scenarios, yields and cropping intensities increase in all regions on both rainfed and irrigated land. The yield and cropping intensity trends under each practice are the same in the two scenarios. In the *Policy Reform* scenario, lower chemical inputs are compensated by improved practices (see Chapter 5). Thus, lower yields in some regions in *Policy Reform* are due to less reliance on irrigation than in *Market Forces* (Annex F-7) which, in turn, is driven by policies to reduce water stress. The higher crop production requirements of *Policy Reform* (Annex F-4) are met by greater expansion of cropped area.

F-6: 1995 figures based on FAO (1999b). In scenarios, requirements for cropland are driven by crop requirements (Annex F-4), moderated by increases in yields (Annex F-5). Increases in crop area are limited by the available potential cultivable land (Annex F-8), which places limits on domestic crop production (Annex F-4). Cropland shrinks in some regions, particularly industrialized regions, as growing output is more than offset by productivity increases. Cropland growth is rapid in Africa, to meet rising demand in the region; and in Latin America, which continues to be a significant exporter throughout the time frame of the scenarios.

F-7: 1995 figures based on FAO (1999b). The roughly 20 percent increase in irrigated land in the scenarios by the year 2050 is less than historical rates of increase. The exhaustion of appropriate and economic new sites, and water constraints limit the expansion. In the Middle East, the sustainability target for limiting increases in water stress is met in part through controlling expansion of irrigated land (see Chapter 5). Higher crop requirements in *Policy Reform* compared to *Market Forces*, combined with limitations on expanding irrigated area, leads to a larger area of rainfed cropland. This is reflected in the smaller fraction of cropland irrigated in *Policy Reform* for some regions.

F-8: PCL based on data from FAO/IIASA (2000). Potential cultivable land includes active farmland and other land-use types potentially suitable for farming. It is lost through land degradation and conversion to the built environment (see Table 4.1). Thus, the area of potential cultivable land in the *Policy Reform* scenario is higher than in *Market Forces*, due in part to less conversion to the built environment, and in part to restoration of degraded cropland over the course of the scenario (see Chapter 5). The severe land constraints seen in the Middle East and South Asia limit the expansion of crop production in those regions, causing them to rely increasingly on imports in both scenarios (Annex F-4).

Environmental pressure

P-1: Total freshwater withdrawals for 1995 from Gleick (1998). In the *Policy Reform* scenario, the sustainability goal of limiting water stress constrains regional withdrawals, particularly in water-scarce regions. In water-abundant regions, ecosystem preservation goals are reflected in part in efficiency improvements relative to *Market Forces* practices, which contribute to lower water requirements in the *Policy Reform* scenario. In water-scarce areas, the potential for improving water-use efficiency limits the degree to which water stress can be reduced (see Chapter 5; Kemp-Benedict *et al.* 2002).

P-2: In the *Market Forces* scenario, strong industrial growth, combined with limited expansion of irrigated land (Annex F-7), leads to growth in the share of industrial water use in total withdrawals. In the *Policy Reform* scenario, rapid income growth in developing regions drives domestic water use upward, increasing its global share. Exploiting the considerable potential for reducing industrial withdrawals shrinks industry's share of total withdrawals over the scenario period (see Chapter 5), while irrigation water requirements remain close to 1995 levels, despite increases in irrigated crop production due to region-specific increases in irrigation efficiency. *Policy Reform* water use is 40 percent below *Market Forces* levels in 2050, due primarily to large improvements in agricultural and industrial efficiencies.

P-3: Water stress increases with increasing pressure on renewable water supplies, as measured by the Use/Resource ratio. The Use/Resource ratio is given by total freshwater requirements divided by renewable freshwater resources, including river flows from adjacent countries. Water supplied from desalinization plants and recycled wastewater is subtracted from total requirements before computing the ratio. Base year data on requirements and supply are taken from Gleick (1998). Data on desalinization and wastewater based on FAO (1996a). The reported water stress figures are regional aggregates of country-level estimates (see Chapter 4). At the national level, water stress rises with the Use/Resource ratio. When the Use/Resource ratio is less than 0.1, there is no stress; when it reaches 0.4, 90 percent of the population is in water stress; and when the Use/Resource ratio is 1.0, 100 percent of the population is in stress. The sustainability goal of limiting increases in water stress constrains the development of the water scenario in *Policy Reform*. In Africa, South Asia and the Middle East, where a number of countries are water-short, water stress rises in both the *Policy Reform* and *Market Forces* scenarios due to growing populations, rising income and expanding water requirements.

P-4: Emissions are from combustion of fossil fuel only; they exclude net emissions from land-use changes and from nonenergy industrial processes. Emissions are estimated from energy consumption data and emissions factors based on IPCC (1995b). Emissions are reported in giga-tonnes of carbon (GtC), where 1 GtC = 1 billion tonnes carbon. (See page 198 for conversion factors.) The sustainability goal of climate protection limits carbon emissions in the *Policy Reform* scenario. The sustainability goal is translated into a cumulative global emissions target for the 1990–2100 period (Chapter 3). In the *Market Forces* scenario, emissions broadly follow primary energy trends, reflecting minimal "decarbonization" of primary energy supply. In the *Policy Reform* scenario, carbon emissions first increase, from 1995–2025, and then decrease, from 2025–2050. Fuel switching to less carbon-intensive fuels (Annex En-2) largely decouples changes in carbon emissions from changes in primary energy requirements, which are in any case lower in *Policy Reform* (Annex En-2).

P-5: 1995 data and emission coefficients based on Posch *et al.* (1996) and Kuylenstierna (1998). Emissions include energy-related and industrial SO_x emissions. In the scenarios, sulfur emissions tend to follow the trajectory of carbon emissions, because carbon-intensive fuels such as coal tend also to be sulfur-intensive. However, greater flexibility in abatement technology for sulfur permits a lower growth in sulfur emissions than in carbon emissions. In *Policy Reform*, sulfur emissions decline significantly, due to slower growth in energy use, a shift from carbon-intensive fuels, and more aggressive abatement compared to *Market Forces*.

P-6: 1995 data on production of industrial roundwood (fuelwood is excluded) are taken from FAO (2000b). Total forest area from FAO (1999b). Exploitable forest area and potential production on natural forest based on FAO (1998b). Data for plantation areas and production based on FAO (2000c). Exploitation intensity is the ratio of actual production to potential production on natural forest land. Exploitation intensities for 1995 are estimated from forest and plantation area, wood production per hectare of natural and plantation forest, and total production of industrial roundwood.

Note that total wood requirements are only partially met by production from forests. Other sources are: recycled materials, nonwood fiber, and wood from trees outside of forests and plantations. Total requirements for fiber are similar in the two scenarios, but considerably more recycling and use of nonwood fiber leads to lower production from forests and less plantation expansion in the *Policy Reform* scenario compared to the *Market Forces* scenario.

Note that some new plantations come from conversion of natural forest. Therefore, the area of natural forest available for forestry production declines over the course of the scenario due to plantation expansion. This contributes to the loss of natural forest seen on Sheets P-7 and P-8. In the *Policy Reform* scenario, there is less conversion of natural forest to plantations to help meet ecosystem preservation goals than in the *Market Forces* scenario.

P-7:   Detailed land-use categories used elsewhere in this report (see Figure 4.10 and Annex P-8) are combined into four aggregate categories: built environment, agricultural land (cropland and grazing land), forest (natural and plantation), and all other land. In the aggregate, patterns are very similar in the two scenarios, and changes over the scenarios are relatively small when expressed as shares of total land area. However, small changes in global land-use shares can correspond to significant absolute changes in area, shown on Annex P-8.

P-8:   1995 areas under cropland, forest and pasture from FAO (1999b). Areas under built environment for developing regions based on Fischer (1993); values for other regions estimated from various sources (see Table 4.3). Protected areas from FAO (1998b) and WCMC (1998b). Plantation areas from FAO (2000c). Area of "Other land" computed as a balance using total land area from FAO (1999b).

The global pattern of change in the *Market Forces* scenario is dominated by expansion of built land and agricultural land at the expense of forest and "Other" land. Changes are comparatively modest in OECD regions, where the major shift is from natural forest to forest plantations. In *Policy Reform*, the global pattern depends on the protection of forests and other ecosystems reflected in the sustainability goals. As part of meeting the goals, there is less expansion of the built environment, agricultural land, and plantation land, compared to the *Market Forces* scenario. Furthermore, most new land under these categories comes from "Other" land in the *Policy Reform* scenario, rather than forest.

P-9:   1995 figures for consumption of nitrogen fertilizer are based on crop-specific application rates reported in FAO (1999a). Fertilizer application rates are averages for countries with data, and do not include any fertilizer applied to fodder and pasture. In the *Market Forces* scenario, increasing yields are associated with increasing fertilizer application rates, based on current patterns. The *Policy Reform* scenario assumes the greater penetration of lower-input farming techniques (see Chapter 5).

P-10:   Total fish and seafood production in 1995 based on FAO (1999b); aquaculture production and maximum sustainable fisheries catch taken from FAO (1997c). Maximum sustainable catch in conventional fisheries is higher in the *Policy Reform* scenario than in the *Market Forces* scenario, as better management of capture fisheries is assumed to extend the current production of approximately 85 Mt per year to a sustained annual production of 100 Mt (FAO 1997c).

P-11:   Data for 1995 based on World Bank Industrial Pollution Projection System (IPPS) (Hettige *et al.* 1994). The IPPS emission factors per dollar value added at 3-digit ISIC level are used to construct "lower-bound" estimates of toxic emissions from industry. In the *Market Forces* scenario, emission factors remain at 1995 levels. Changes in emissions in the *Market Forces* scenario, therefore, reflect growth in industrial production and the effect of shifts in industrial composition from highly polluting industries to less-polluting ones. *Policy Reform* toxic emission intensities are reduced by a factor of roughly 10 by 2050, in accordance with the sustainability target (see Chapter 3).

References

Adriaanse, A., Bringezu, S., Hammond, A., Moriguchi, Y., Rodenburg, E., Rogich, D. and Schutz, H. (1997) *Resource Flows: The Material Basis of Industrial Economies*, World Resources Institute/Wuppertal Institute/Netherlands Ministry of Housing, Spatial Planning and Environment/Japan National Institute for Environmental Studies, Washington, DC: World Resources Institute.

Alcamo, J., Döll, P., Kaspar, F. and Siebert, S. (1997) *Global Change and Global Scenarios of Water Use and Availability: An Application of Water GAP 1.0*, Kassel, Germany: Wissenschaftliches Zentrum für Umweltsystemforschung, Universität Gesamthochschule Kassel.

Alexandratos, N. (ed.) (1995) *World Agriculture: Towards 2010*, Chichester, UK: John Wiley and Sons.

Azar, C. and Rodhe, H. (1997) "Targets for Stabilization of Atmospheric CO_2," *Science*, 276: 92–93.

Barber, B. (1995) *Jihad vs. McWorld*, New York: Random House.

Barney, G.O. (1980) *The Global 2000 Report to the President: Entering the Twenty-First Century: A Report*, Washington, DC: US Government Printing Office.

Barney, G.O. (1993) *Global 2000 Revisited: What shall we do? The Critical Issues of the 21st Century*, Arlington, VA: Millennium Institute.

Beckerman, W. (1995) *Small is Stupid*, London: Duckworth.

Bender, J. (1994) *Future Harvest: Pesticide-Free Farming*, Lincoln: University of Nebraska Press.

Bernardini, O. and Galli, R. (1993) "Dematerialization: Long-term Trends in the Intensity of Use of Materials and Energy," *Futures*, 25: 431–448.

Board on Sustainable Development of the US National Research Council (BSD) (1998) *Our Common Journey: Navigating a Sustainability Transition*, Washington, DC: National Academy Press.

Bongaarts, J. (1997) Personal communication.

Bossel, H. (1998) *Earth at a Crossroads: Paths to a Sustainable Future*, Cambridge, UK: Cambridge University Press.

Brinkhoff, T. (2001) *City Population*, Online. Available HTTP: <http://www.citypopulation.de> (accessed 17 October 2001).

Brown, L. (1978) *The Twenty-Ninth Day: Accommodating Human Needs and Numbers to the Earth's Resources*, New York: W.W. Norton & Company.

Brown, L.R., Flavin, C., French, H., Abramowitz, J., Bright, C., Gardner, G., McGinn, A., Renner, M., Roodman, D. and Starke, L. (1997) *State of the World: A Worldwatch Institute Report on Progress Toward a Sustainable Society*, New York: W.W. Norton & Company.

Brown, L., Renner, M. and Flavin, C. (1998) *Vital Signs 1998*, New York: W.W. Norton & Company.

Bryant, D., Nielsen, D. and Tangley, L. (1997) *The Last Frontier Forests: Ecosystems and Economies on the Edge*, Washington, DC: World Resources Institute.

Burke, L. and Bryant, D. (1998) *Reefs at Risk*, Washington, DC: World Resources Institute.

Burrows, B., Mayne, A. and Newbury, P. (1991) *Into the 21st Century: A Handbook for a Sustainable Future*, Twichenham, England: Adamantine Press.

Campbell, C.J. (2001) *Evolution of Oil Assessments*, Online. Available HTTP: <http://www.hubbertpeak.com/campbell/assessments.htm> (accessed 15 October 2001).

Carley, M. and Spapens, P. (1998) *Sharing the World: Sustainable Living and Global Equity in the 21st Century*, London: Earthscan.

Central Planning Bureau (The Netherlands) (1992) *Scanning the Future: A Long-term Scenario Study of the World Economy 1990–2015*, The Hague: SDC Publishers.

Chadwick, M. (1994) "Visions of a Sustainable World: Ethical Evaluations or Political Programmes?" in F. Graham-Smith (ed.) *Population, the Complex Reality*, London: The Royal Society.

Cole, S. (1981) "Methods of analysis for long-term development issues," in UNESCO *Methods for Development Planning*, Paris: UNESCO Press.

Consultative Group on Sustainable Development Indicators (CGSDI) (2001) Online. Available HTTP: <http://iisd1.iisd.ca/cgsdi/default.htm> (accessed 22 October 2001).

Conway, G. (1997) *The Doubly Green Revolution: Food for All in the 21st Century*, London: Penguin Books.

Daily, G. (1997) *Nature's Services: Societal Dependence on Natural Ecosystems*, Washington, DC: Island Press.

Daly, H. (1996) *Beyond Growth: The Economics of Sustainable Development*, Boston: Beacon Press.

Deffeyes, K. (2001). *Hubbert's Peak: The Impending World Oil Shortage*, Princeton, New Jersey: Princeton University Press.

de Haan, C., Steinfeld, H. and Blackburn, H. (1996) *Livestock and the Environment: Finding a Balance*, Suffolk, UK: WRENmedia. Online. Available HTTP: <http://www.fao.org>.

Deininger, K. and Squire, L. (1996) "A New Data Set Measuring Income Inequality," *The World Bank Economic Review*, 10(3): 565–591.

Dethlefson, V., Jackson, T. and Taylor, P. (1993) "The Precautionary Principle," in T. Jackson (ed.) *Clean Production Strategies: Developing Preventive Environmental Maintenance in the Industrial Economy*, Boca Raton, Florida: Lewis Publishers.

Douglas, I. (1994) "Human Settlements," in W.B. Meyer and B.L. Turner (eds) *Changes in Land Use and Land Cover: A Global Perspective*, Cambridge: Cambridge University Press.

Durning, A.B. and Brough, H.B. (1991) *Taking Stock: Animal Farming and the Environment*, Worldwatch Paper 103, Washington, DC: Worldwatch Institute.

Earth Charter Initiative (ECI) (2000) *The Earth Charter*, San José, Costa Rica: Earth Charter Commission Secretariat. Online. Available HTTP: <http://www.earthcharter.org/draft/charter.rtf> (accessed 29 November 2001).

Ehrlich, P. (1968) *The Population Bomb*, New York: Ballantine Books.

Ehrlich, P. and Ehrlich, A. (1990) *The Population Explosion*, New York: Simon and Schuster.

Energy Innovations (1997) *Energy Innovations: A Prosperous Path to a Clean Environment*, Washington, DC: Alliance to Save Energy, American Council for an Energy-Efficient Economy, Natural Resources Defense Council, Tellus Institute, and Union of Concerned Scientists.

Factor 10 Club (1995) *Carnoules Declaration*, Wuppertal: Wuppertal Institut.

Fischer, G. (1993) Personal communication.

Flavin, C. (1984) "Reassessing the Economics of Nuclear Power," in State of the World, New York: W.W. Norton & Company.

Food and Agriculture Organization (FAO) (1996a) *AQUASTAT Database*, Rome: Food and Agriculture Organization. Online. Available HTTP: <http://www.fao.org> (accessed 20 March 1998).

—— (1996b) *Report of the World Food Summit*, Rome: Food and Agriculture Organization.

—— (1996c) *Sixth World Food Survey*, Rome: Food and Agriculture Organization.

—— (1997a) *Mapping Undernutrition* (poster), Rome: Food and Agriculture Organization.

—— (1997b) *State of the World's Forests 1997*, Rome: Food and Agriculture Organization.

—— (1997c) *The State of World Fisheries and Aquaculture 1996*, Rome: Food and Agriculture Organization.

—— (1998a) *Fishstat Plus ver. 2.19* (database), Rome: Food and Agriculture Organization.

—— (1998b) *Global Fibre Supply Model*, FAO Forestry Department, Forest Products Division, Wood and Non-Wood Products Utilization Branch, Rome: FAO.

—— (1999a) *Fertilizer Use by Crop 1999*, Rome: Food and Agriculture Organization. Microsoft Excel workbook, Online. Available HTTP: <http://www.fao.org> (accessed 3 July 2000).

—— (1999b) *FAOSTAT 98 CD-ROM*, Rome: Food and Agriculture Organization of the United Nations.

—— (1999c) *State of Food and Agriculture 1999*, Rome: Food and Agriculture Organization of the United Nations.

—— (1999d) *State of the World's Forests 1999*, Rome: Food and Agriculture Organization.

—— (2000a) *Agriculture: Towards 2015/30, Technical Interim Report, April 2000*, Rome: Food and Agriculture Organization of the United Nations.

—— (2000b) *FAOSTAT Online Database*, Rome: Food and Agriculture Organization. Online. Available HTTP: <http://apps.fao.org/> (accessed 5 July 2000).

—— (2000c) *The Global Outlook for Future Wood Supply from Forest Plantations*, Global Forest Products Outlook Study Working Paper GFPOS/WP/03, Rome: FAO.

—— (2000d) *Land Resource Potential and Constraints at Regional and Country Levels*, World Soil Resources Report 90, Rome: FAO.

—— (2000e) *The State of Food Insecurity in the World 2000*, Rome: Food and Agriculture Organization of the United Nations.

—— (2000f) *The State of World Fisheries and Aquaculture 2000*. Rome: Food and Agriculture Organization.

—— (2000g) *Tropical Feeds Database ver. 8.0*, Rome: Food and Agriculture Organization. Online. Available HTTP: <http://www.fao.org/> (accessed 13 June 2000).

Food and Agriculture Organization of the United Nations/International Institute for Applied Systems Analysis (FAO/IIASA) (2000) *Global Agro-ecological Zones*, Online. Available HTTP: <http://www.iiasa.ac.at/Research/LUC/GAEZ/index.htm> (accessed 9 August 2000).

Forest Resources Assessment 1990 Project (FRA) (1996) *Forest Resources Assessment 1990: Survey of Tropical Forest Cover and Study of Change Processes*, Rome: Food and Agriculture Organization of the United Nations.

Gallopín, G.C. (1991) "Human Dimensions of Global Change: Linking the Global and the Local Processes," *International Social Science Journal*, 130: 707–718.

—— (1994) *Impoverishment and Sustainable Development, A Systems Approach*, Winnipeg: International Institute for Sustainable Development.

Gallopín, G., Gutman, P. and Maletta, H. (1989) "Global Impoverishment, Sustainable Development and the Environment: A Conceptual Approach," *International Social Science Journal*, 121: 375–397.

Gallopín, G., Hammond, A., Raskin, P. and Swart, R. (1997) *Branch Points: Global Scenarios and Human Choice*, PoleStar Series Report No. 7, Stockholm, Sweden: Stockholm Environment Institute.

Gandhi, M. (1993) *The Essential Writings of Mahatma Gandhi*, New York: Oxford University Press.

Gleick, P. (1998) *The World's Water 1998–1999: The Biennial Report on Freshwater Resources*, Washington, DC: Island Press.

—— (2000) *The World's Water: The Biennial Report on Freshwater Resources 2000–2001*, Washington, DC: Island Press.

Global Scenario Group (GSG) (2001) Online. Available HTTP: <http://www.gsg.org/> (accessed 5 December 2001).

Goodland, R., Daly, H. and El Serafy, S. (1992) *Population, Technology, and Lifestyle*, Washington, DC: Island Press.

Gurr, T.R. (1968) "A Causal Model of Civil Strife: A Comparative Analysis Using New Indices," *American Political Science Review*, 62: 1104–1124.

Hare, W. (1997) *The Carbon Logic*, Amsterdam: Greenpeace.

Herrera, A.D., Scolnic, H., Chichilnisky, G., Gallopín, G., Hardoy, J., Mosovich, D., Oteiza, E., de Romero Brest, G., Suarez, C. and Talavera, L. (1976) *Catastrophe or New Society? A Latin American World Model*, Ottawa: International Development Research Centre.

Heston, A., Summers, R., Nuxoll, D.A. and Aten, B. (1995) *Penn World Table (Mark 5.6a)*. Cambridge, MA: National Bureau of Economic Research.

Hettige, H., Martin, P., Singh, M. and Wheeler, D. (1994) *The Industrial Pollution Projection System*, Washington, DC: The World Bank.

Hobbes, T. (1977, first published in 1651) *The Leviathan*, New York: Penguin Press.

Holling, C.S. (ed.) (1978) *Adaptive Environmental Assessment and Management*, Chichester, UK: John Wiley & Sons.

Holling, C.S. (ed.) (1986) "The Resilience of Terrestrial Ecosystems: Local Surprise and Global Change," in W.W. Clark and R.R. Munn (eds) *Sustainable Development of the Biosphere*, Oxford: Oxford University Press.

Homewood, K.M. and Rogers, W.A. (1991) *Maasailand Ecology: Pastoralist Development and Wildlife Conservation in Ngorongoro, Tanzania*, Cambridge: Cambridge University Press.

Houghton, J.T., Ding, Y., Griggs, D.J., Noguer, M., van der Linden, P.J., Da, X., Maskell, K. and Johnson, C.A. (2001) *Climate Change 2001: The Scientific Basis*, Online. Available HTTP: <http://www.ipcc.ch/pub/tar/wg1/001.htm> (accessed 13 November 2001).

Hubbert, M.K. (1956) "Nuclear Energy and the Fossil Fuels," in *American Petroleum Institute, Proceedings of the Spring Meeting: Drilling and Production Practice*, San Antonio, Texas, July 25.

Hulme, M., Wigley, T.M.L., Barrow, E.M., Raper, S.C.B., Centella, A., Smith, S. and Chipanshi, A.C. (2000) *Using a Climate Scenario Generator for Vulnerability and Adaptation Assessments: MAGICC and SCENGEN Version 2.4 Workbook*, Norwich, UK: Climatic Research Unit.

Intergovernmental Panel on Climate Change (IPCC) (1992) *1992 IPCC Supplement*, Geneva: World Meteorological Organization.

—— (1995a) *Climate Change 1994: Radiative Forcing of Climate Change and An Evaluation of the IPCC IS92 Emission Scenarios*. J.T. Houghton, L.G. Meira Filho, J. Bruce, Hoesung Lee, B.A. Callander, E. Haites, N. Harris and K. Maskell (eds). Cambridge: Cambridge University Press.

—— (1995b) *IPCC Greenhouse Gas Inventory Reference Manual: IPCC Guidelines for National Greenhouse Gas Inventories*. Vol. 3. Bracknell, UK: IPCC.

Intergovernmental Panel on Climate Change/Organization for Economic Cooperation and Development/International Energy Agency (IPCC/OECD/IEA) (1996) *Revised 1996 IPCC Guidelines for National Greenhouse Gas Inventories*, Vol. 3, J.T. Houghton, L.G. Meira Filho, B. Lim, K. Treanton, I. Mamaty, Y. Bonduki, D.J. Griggs and B.A. Callender (eds). Bracknell, UK: UK Meteorological Office.

International Energy Agency (IEA) (1999) *Energy Balances 1999*, Paris: OECD/IEA.

Jackson, T. and McGillivray, A. (1995) *Accounting for Toxic Emissions from the Global Economy: The Case of Cadmium*, PoleStar Series Report No. 6, Stockholm: Stockholm Environment Institute.

Jackson, T. and Taylor, P. (1992) "The Precautionary Principle and the Prevention of Marine Pollution," *Journal of Chemistry and Ecology*, 7: 123–134.

Kahn, H. (1962) *Thinking About the Unthinkable*, New York: Avon Books.

Kahn, H. and Wiener, A. (1967) *The Year 2000*, New York: MacMillan.

Kahn, H., Brown, W. and Martel, L. (1976) *The Next 2000 Years: A Scenario for America and the World*, New York: Morrow.

Kaplan, R. (2000) *The Coming Anarchy*, New York: Random House.

Kates, R.W., Clark, W.C., Corell, R., Hall, J.M., Jaeger, C.C., Lowe, I., McCarthy, J.J., Schellnhuber, H.J., Bolin, B., Dickson, N.M., Faucheux, S., Gallopín, G.C., Grübler, A., Huntley, B., Jäger, J., Jodha, N.S., Kasperson, R.E., Mabogunje, A., Matson, P., Mooney, H., Moore, B., O'Riordan, T. and Svedin, U. (2001) "Sustainability Science," *Science*, 292: 641–642.

Kemp-Benedict, E., Heaps, C. and Raskin, P. (2002) *Global Scenario Group Futures: Technical Notes*, Boston: Stockholm Environment Institute-Boston. Online. Available HTTP: <http://gsg.org/gsgpub.html>.

Kendall, H. and Pimentel, D. (1994) "Constraints on the Expansion of the Global Food Supply," *Ambio*, 23(3): 198–205.

Keynes, J.M. (1936) *The General Theory of Employment, Interest, and Money*, London: MacMillan.

Krause, F., Bach, W. and Koomey, J. (1989) Energy Policy in the Greenhouse. Volume 1: From Warming Fate to Warming Limit: Benchmarks for a Global Climate Convention. European

Environmental Bureau, Brussels/International Project for Sustainable Energy Paths, El Cerrito, California (1990 edition published by Earthscan, London). Cited in: Karas, J.H.W. (1991) *Back From the Brink: Greenhouse Gas Targets for a Sustainable World*, London: Friends of the Earth.

Kuylenstierna, J. (1998) Personal communication.

Kuznets, S. (1967) "Population and Economic Growth," *Proceedings of the American Philosophical Society*, 3: 170–193.

Lalonde, L.-G. and Sukigara, T. (1997) *LDPS² User's Guide*, Rome: Food and Agriculture Organization, Animal Production and Health Division.

Landes, D.S. (1970) *The Unbound Prometheus: Technological Change and Industrial Development in Western Europe from 1750 to the Present*, Cambridge: Cambridge University Press.

Leach, G. (1995) *Global Land and Food in the 21st Century: Trends and Issues for Sustainability*, PoleStar Series Report No. 5, Stockholm: Stockholm Environment Institute.

Leontieff, W. (1976) *The Future of the World Economy: A Study on the Impact of Prospective Economic Issues and Policies on the International Development Strategy*, New York: United Nations.

Lonergan, S. and Brooks, D. (1994) *Watershed: The Role of Fresh Water in the Israeli–Palestinian Conflict*, Ottawa: International Development Research Centre.

Lovins, A. and Price, J. (1975) *Non-Nuclear Futures*, Cambridge, MA.: Ballinger Publishing Co.

McCarthy, J.J., Canziani, O.F., Leary, N.A., Dokken, D.J. and White, K.S. (2001) *Climate Change 2001: Impacts, Adaptation, and Vulnerability*, Online. Available HTTP: <http://www.ipcc.ch/pub/tar/wg2/001.htm> (accessed 13 November 2001).

Mackenzie J.J. (2000) *Oil as a Finite Resource: When is Global Production Likely to Peak?* Washington, DC: World Resources Institute. Online. Available HTTP: <http://www.wri.org/wri/climate/jm_oil_000.html> (accessed 10 December 2001)

Malthus, T. (1983, first published 1798) *An Essay on the Principle of Population*, New York: Penguin Press.

Matthews, E. and Hammond, A. (1999) *Critical Consumption Trends and Implications: Degrading Earth's Ecosystems*, Washington, DC: World Resources Institute.

Meadows, D.H., Meadows, D.L., Randers, J. and Behrens, W.W. (1972) *Limits to Growth*, New York: Universe Books.

Meadows, D.H., Meadows, D.L. and Randers, J. (1992) *Beyond the Limits: Confronting Global Collapse, Envisioning a Sustainable Future*, White River Junction, Vermont: Chelsea Green Publishing Company.

Meadows, D.L , Richardson, J. and Bruckmann, G. (1982) *Groping in the Dark: The First Decade of Global Modeling*, New York: John Wiley & Sons.

Mesarovic, M.D. and Pestel, E. (1974) *Mankind at a Turning Point*, New York: Dutton.

Milbrath, L.W. (1989) *Envisioning a Sustainable Society: Learning Our Way Out*, Albany, New York: SUNY Press.

Miles, I. (1981) "Scenario Analysis: Identifying Ideologies and Issues," in UNESCO, *Methods for Development Planning*, Paris: UNESCO Press.

Mill, J.S. (1998, first published 1848) *Principles of Political Economy*, Oxford: Oxford University Press.

Moldan, B., Billharz, S. and Matravers, R. (1997) *Sustainability Indicators: A Report on the Project on Indicators of Sustainable Development*, SCOPE 58, Chichester, UK: John Wiley & Sons.

Morita, T., Kainuma, M., Harasawa, H., Kai, K. and Matsuoka, Y. (1995) *Long-term Global Scenarios Based on the AIM Model*, AIM Interim Paper IP-95-03, Nagoya, Japan: National Institute for Environmental Studies.

Muller, E.N. (1988) "Democracy, Economic Development and Income Inequality," *American Sociological Review*, 53: 50–68.

Muller, E.N. and Seligson, M.A. (1987) "Inequality and Insurgency," *American Political Science Review*, 81: 425–451.

Munasinghe, M. and Shearer, W. (eds) (1995) *Defining and Measuring Sustainability*, Washington, DC: United Nations University and The World Bank.

Munn, T., Whyte, A. and Timmerman, P. (1999) "Emerging Environmental Issues: A Global Perspective of SCOPE," *Ambio*, 28 (6): 464–471.

Myers, N. (ed.) (1984) *Gaia: An Atlas of Planet Management*, New York: Anchor Press Doubleday.

Nakićenović, N., Alcamo, J., Davis, G., de Vries, B., Fenhann, J., Gaffin, S., Gregory, K., Grübler, A., Jung, T.Y., Kram, T., Lebre La Rovere, E., Michaelis, L., Mori, S., Morita, T., Pepper, W., Pitcher, H., Price, L., Riahi, K., Roehrl, A., Rogner, H.-H., Sankovski, A., Schlesinger, M., Shukla, P., Smith, S., Swart, R., van Rooijen, S., Victor, N. and Dadi, Z. (2000) *Special Report on Emissions Scenarios: A Special Report of Working Group III of the Intergovernmental Panel on Climate Change*, Cambridge: Cambridge University Press.

Oldeman, L., Hakkeling, R. and Sombroek, W. (1991) *World Map of the Status of Human-Induced Soil Degradation: An Explanatory Note*, Second revised edition. Nairobi: United Nations Environment Program.

Organisation for Economic Co-operation and Development (OECD) (1997) *The World in 2020: Towards a New Global Age*, Paris: OECD.

—— (1998) *Environmental Outlook*, Paris: OECD.

—— (2001) *Environmental Outlook*, Paris: OECD.

Paddock, W. and Paddock, P. (1967) *Famine, 1975! America's Decision: Who Will Survive?* Boston: Little, Brown.

Pinstrup-Anderson, P., Pandya-Lorch, R. and Rosegrant, M.W. (1997) *The World Food Situation: Recent Developments, Emerging Issues, and Long-Term Prospects*, 2020 Vision Food Policy Report, Washington, DC: International Food Policy Research Institute.

Posch, M., Hettelingh, J.-P., Alcamo, J. and Krol, M. (1996) "Integrated Scenarios of Acidification and Climate Change in Asia and Europe," *Global Environmental Change*, 6(4): 375–394.

Postel, S. (1992) *Last Oasis: Facing Water Scarcity*, New York: W.W. Norton & Company.

Postel, S., Daily, G. and Ehrlich, P. (1996) "Human Appropriation of Renewable Freshwater," *Science*, 271: 785–788.

Prahad, C.K. and Hart, S.L. (1999) *Strategies for the Bottom of the Pyramid: Creating Sustainable Development*, Online. Available HTTP: <http://www.igc.apc.org/wri/meb/wrisummit/resource.html> (accessed 27 November 2001).

Raskin, P. (1995) "Methods for Estimating the Population Contribution to Environmental Change," *Ecological Economics*, 15: 225–233.

Raskin, P. and Margolis, R. (1995) *Global Energy in the 21st Century: Patterns, Projections and Problems*, PoleStar Series Report No. 3, Stockholm, Sweden: Stockholm Environment Institute.

Raskin, P., Hansen, E. and Margolis, R. (1995) *Water and Sustainability: A Global Outlook*, PoleStar Series Report No. 4, Stockholm, Sweden: Stockholm Environment Institute.

Raskin, P., Chadwick, M., Jackson, T. and Leach, G. (1996) *The Sustainability Transition: Beyond Conventional Development*, PoleStar Series Report No. 1, Stockholm, Sweden: Stockholm Environment Institute.

Raskin, P., Gleick, P., Kirshen, P., Pontius, G. and Strzepek, K. (1997) *Water Futures: Assessment of Long-range Patterns and Problems*, Background Document for the SEI/United Nations Comprehensive Assessment of the Freshwater Resources of the World, Stockholm: Stockholm Environment Institute.

Raskin, P. and Margolis, R. (1998) "Global Energy, Sustainability, and the Conventional Development Paradigm," *Energy Sources*, 20: 363–383.

Raskin, P., Gallopín, G., Gutman, P., Hammond, A. and Swart, R. (1998) *Bending the Curve: Toward Global Sustainability: A Report of the Global Scenario Group*, PoleStar Series Report No. 8, Stockholm, Sweden: Stockholm Environment Institute.

Raskin, P., Heaps, C., Sieber, J., and Kemp-Benedict, E. (1999) *Polestar System Manual*, Boston: Stockholm Environment Institute-Boston. Online. Available HTTP: <http://www.tellus.org/seib/publications/ps2000.pdf> (accessed 10 December 2001).

Raskin, P. and Kemp-Benedict, E. (2002) *GEO Scenario Framework*. Nairobi: United Nations Environment Program.

Raskin, P., Banuri, T., Gallopín, G., Gutman, P., Hammond, A., Kates, R. and Swart, R. (2002) *Great Transition: The Promise and Lure of the Times Ahead*, PoleStar Series Report No. 10, Stockholm, Sweden: Stockholm Environment Institute.

Ravallion, M., Datt, G. and van de Walle, D. (1991) "Quantifying Absolute Poverty in the Developing World," *Review of Income and Wealth*, 37(4): 345–361.

Rijsberman, F. and Swart, R. (eds) (1990) *Targets and Indicators of Climate Change*, Stockholm, Sweden: Stockholm Environment Institute.

Robinson, J. (1990) "Futures Under Glass: A Recipe for People Who Hate to Predict," *Futures*, 22: 820–841.

Rose, D., Basiotis, P.P. and Klein, B.W. (1995) "Improving Federal Efforts to Assess Hunger and Food Insecurity," *Food Review*, 18(1): 18–23.

Rosegrant, M., Agacaoili-Sobilla, M. and Perez, N. (1995) *Global Food Projections to 2020: Implications for Investment*, Washington, DC: International Food Policy Research Institute.

Ruiz-Huerta, J., Martinez, R. and Ayala, L. (1999) "Earnings Inequality, Unemployment and Income Distribution in the OECD," *Luxembourg Income Study Working Paper* No. 214, Syracuse, New York: Maxwell School of Citizenship and Public Affairs, Syracuse University.

Sales, K. (2000) *Dwellers in the Land: The Bioregional Vision*, Athens, GA: University of Georgia Press.

Scherr, S. and Yadav, S. (1996) *Land Degradation in the Developing World: Implications for Food, Agriculture, and the Environment to 2050*, Washington, DC: International Food Policy Research Institute.

Schumacher, E.F. (1972) *Small is Beautiful*, London: Blond and Briggs.

Schwartz, P. (1991) *The Art of the Long View*, New York: Currency Doubleday.

Schwartz, P. and Leyden, P. (1997) "The Long Boom: A History of the Future," *Wired*, 5: 115–129.

Seckler, D., Amarasinghe, U., Radhika de Silva, D. and Barker, R. (1998) *World Water Demand and Supply, 1990 to 2025: Scenarios and Issues*, IIMI Research Report 19, Colombo, Sri Lanka: International Water Management Institute.

Sen, A. (1981) *Poverty and Famines: An Essay on Entitlement and Deprivation*, Oxford: Clarendon Press.

Shaw, R. Gallopín, G.C., Weaver, P. and Oberg, S. (1991) *Sustainable Development: A Systems Approach*, Laxenberg, Austria: International Institute for Applied Systems Analysis.

Shiklomanov, I. (1997) *Assessment of Water Resources and Water Availability of the World*, Geneva: World Meteorological Organization.

Simon, J. (1981) *The Ultimate Resource*, Princeton, New Jersey: Princeton University Press.

Smith, A. (1991, first published in 1776) *The Wealth of Nations*, Amherst, New York: Prometheus Books.

Steenbergen, van B. (1994) "Global Modelling in the 1990's," *Futures*, 26(1): 44–56.

Stockholm Environment Institute (SEI) (1997) *Comprehensive Assessment of the Freshwater Resources of the World*, Stockholm, Sweden: Stockholm Environment Institute.

Svedin, U. and Aniansson, B. (eds) (1987) *Surprising Futures: Notes from an International Workshop on Long-term World Development*, Stockholm, Sweden: Swedish Council for Planning and Coordination of Research.

Swallow, B.M. (1999) *Impacts of Trypanosomiasis on African Agriculture*, Nairobi: International Livestock Research Institute.

Swaminathan, M. (1997) "Sustainable Development: Five Years after Rio," in *A Better Future for Planet Earth*, Tokyo: Asahi Glass Foundation.

Swart, R. (1996) "Security Risks of Global Environmental Changes: Viewpoint," *Global Environmental Change*, 3(3): 187–192.

Tabatabai, H. (1996) *Statistics on Poverty and Income Distribution: An ILO Compendium of Data*, Geneva: International Labour Office.

Thompson, P. (1993) *The Work of William Morris*, Oxford: Oxford University Press.

Thurow, L.C. (1996) *The Future of Capitalism: How Today's Economic Forces Shape Tomorrow's World*, New York: Penguin Books.

Toth, F.I., Hizsnyik, E. and Clark, W. (eds) (1989) *Scenarios of Socio-economic Development for Studies of Global Environmental Change: A Critical Review*, Laxenburg, Austria: International Institute for Applied Systems Analysis.

United Nations (UN) (1999) Annual *Populations 1950–2050: The 1998 Revision*, New York: United Nations.

United Nations Conference on Environment and Development (UNCED) (1992) *Agenda 21: Programme of Action for Sustainable Development*, New York: United Nations.

United Nations Department for Policy Coordination and Sustainable Development (UNDPCSD) (1997) *Critical Trends: Global Change and Sustainable Development*, New York: United Nations.

United Nations Development Programme (UNDP) (1997) *Human Development Report 1997*, Oxford: Oxford University Press.

—— (2001) *Human Development Report 2001*, Oxford: Oxford University Press.

United Nations Educational, Scientific and Cultural Organization (UNESCO) (1997) *Fifth International Conference on Adult Education*, Paris: United Nations Educational, Scientific and Cultural Organization.

United Nations Environment Programme (UNEP) (1997) *Global Environment Outlook*, Nairobi, Kenya: United Nations Environment Program.

United Nations Framework Convention on Climate Change (UNFCCC) (1997) *Kyoto Protocol to the United Nations Framework Convention on Climate Change*, Document FCCC/CP/1997/7/Add.1, Online. Available HTTP: <http://www.unfccc.de> (accessed 20 December 2001).

United Nations University/World Institute for Development Economics Research (UNU/WIDER) (1999) *UNDP World Income Inequality Database*, Online. Available HTTP: <http://www.wider.unu.edu/wiid/wiid.htm> (accessed 28 August 2000).

United States Bureau of the Census (USBC) (1997) *Historical Income and Poverty Tables (Table H-4)*, Washington, DC: United States Census Bureau. Online. Available HTTP: <http://www.census.gov/hhes/income/histinc/index/html> (accessed 20 October 1997).

—— (2000) *US Census 2000*, Washington, DC: United States Census Bureau.

United States Department of Agriculture (USDA) (1994) *1992 National Resources Inventory: 1992 NRI Changes in Land Cover/Use Between 1982 and 1992*, Online. Available HTTP: <http://www.nhq.nrcs.usda.gov/land/tables/t2211.html>.

United States Department of Commerce, Bureau of the Census (USBC) (1992) *Statistical Abstract of the United States*. Washington, DC: US Government Publishing Office.

—— (1995) *Statistical Abstract of the United States*. Washington, DC: US Government Publishing Office.

United States Geological Survey (USGS) (2000) *World Petroleum Assessment 2000: Description and Results*, Online. Available HTTP: <http://greenwood.cr.usgs.gov/energy/WorldEnergy/DDS-60/> (accessed 15 October 2001).

United States Nuclear Regulatory Commission (1975) *1975 Reactor Safety Study: An Assessment of Accident Risks in U.S. Commercial Nuclear Power Plants*, Report number WASH-1400, NUREG-75/014 (the "Rasmussen Report"), Washington, DC: U.S. Nuclear Regulatory Commission.

University of Sussex Science Policy Research Unit (1973) *Thinking About the Future: A Critique of the Limits to Growth*, London: Chatto and Windows.

von Asselt, M. and Rotmans, J. (1997) *Uncertainty in Integrated Assessment Modeling: A Cultural Perspective-based Approach*, Bilthoven, The Netherlands: RIVM.

von Hippel, F. (1977) "Looking back on the Rasmussen Report," *Bulletin of the Atomic Scientists*, 33: 2.

Wallensteen, P. and Swain, A. (1997) *International Fresh Water Resources: Conflict or Cooperation?* Background Document for the SEI/United Nations Comprehensive Assessment of the Freshwater Resources of the World, Stockholm, Sweden: Stockholm Environment Institute.

Wang, T.Y. (1993) "Inequality and Political Violence Revisited," *American Political Science Review*, 87(4): 979–983.

Williams, R., Larson, E. and Ross, M. (1987) "Materials, Affluence and Energy Use," *Annual Review of Energy*, 12: 99–144.

World Bank (1990) *World Development Report 1990: Poverty*, New York: Oxford University Press.

—— (1991) *World Development Report 1991*, New York: Oxford University Press. Cited in IPCC (1992) *1992 IPCC Supplement*. Geneva: World Meteorological Organization.

—— (1996) *Towards Environmentally Sustainable Development in Sub-Saharan Africa: A World Bank Agenda*, Washington, DC: World Bank.

—— (1997) *World Development Indicator* (electronic distribution), Washington, DC: World Bank.

—— (1999) Privately communicated data incorporated in World Resources Institute/Stockholm Environment Insitute-Boston (2000) *Is Growth Enough? A Scenario Analysis Prepared for the World Bank*, Washington, DC: World Resources Institute.

—— (2000) *World Development Indicators 2000*, Washington, DC: World Bank.

—— (2001) *World Development Report 2000/2001: Attacking Poverty*, New York: Oxford University Press.

World Business Council for Sustainable Development (WBCSD) (1997) *Exploring Sustainable Development*, Geneva: WBCSD.

World Commission on Dams (WCD) (2000) *Dams and Development: A New Framework for Decision-Making*, London: Earthscan.

World Commission on Environment and Development (WCED) (1987) *Our Common Future*, Oxford: Oxford University Press.

World Conservation Monitoring Centre (WCMC) (1998b) *Global Protected Areas Summary Statistics*. Cambridge, UK: WCMC. Online. Available HTTP: <http://www.wcmc.org.uk> (accessed 10 March 1998).

World Energy Council (WEC) (1992) *1992 Survey of Energy Resources*, London: World Energy Council.

—— (1995) *1995 Survey of Energy Resources*, London: World Energy Council.

World Energy Council/International Institute for Applied Systems Analysis (WEC/IIASA) (1995) *Global Energy Perspectives to 2050 and Beyond*, London: World Energy Council.

World Health Organization (WHO) (1997a) *Health for All in the 21st Century*, Geneva: World Health Organization.

—— (1997b) *The World Health Report 1997: Conquering Suffering, Enriching Humanity*, Geneva: World Health Organization.

World Resources Institute (WRI) (1994) *World Resources, 1994–95*, New York: Oxford University Press.

—— (1996a) *World Resources, 1996–97*, New York: Oxford University Press.

—— (1996b) *World Resources 1996–97 Database Diskette*, Washington, DC: World Resources Institute.

—— (1998) *World Resources, 1998–99*, Washington, DC: World Resources Institute.

World Summit for Social Development (WSSD) (1995) *The Copenhagen Declaration and Programme of Action*, New York: United Nations.

World Trade Organization (WTO) (2001) *International Trade Statistics 2001*, Geneva: WTO Publications.

World Water Vision Commission Report (WWV) (2000) *A Water Secure World: Vision for Water, Life, and the Environment*, Paris: World Water Council.

Wright, R. (2000) *Non Zero: The Logic of Human Destiny*, New York: Pantheon.

Index

Conventional Worlds scenarios (*Continued*)
 Forces variant 23; market orientation of 81;
 Policy Reform variant 23
core global indicators and targets 34
critical trends 126; and driving forces 15–22
crop: requirements and production *Market
 Forces* and *Policy Reform* scenarios 164–5;
 yields 58
cropland 96
cultural and spiritual groups 123
cultural homogenization 78
cultural interchange 50
cultural preferences 124

Decade of Safe Drinking Water 37
decentralization of authority 22
deforestation 48–9, 72
dematerialization 45; of growing economies 103
demographic transition 107; toward stable
 populations 82
demographics 16–17
Denmark 84
desertification 48
destitution and oppression 109
Developing macro-regions 33
development: agenda, revised 116; convergence
 85; failure of 130; paradigm of consumerism
 125; trajectory, future 10
digital divide 20
domestic water intensities 100
doubly green revolution 94
driving forces 6, 12
Dutch Central Planning Bureau 13

Earth Summit 8, 30
Earth's carrying capacity 16
Eastern Europe 85
Eco-communalism 115
eco-efficiency ratio 45
eco-efficient-industrial system 45
economic bargain 117–18
economic efficiency 124; principle of 3
economic equity, dimensions of 82
economic globalization 50, 60
economic growth 32, 91; centrality of 125
economic inequality 19
economic polarization 77
economic system 6
economics 17–18
ecosystem pressure 39, 48–9
eco-taxes 103
education, adult illiteracy 35
Ehrlich, P. 16, 28
electricity 54; world production 63
electronic media 18
emergence of new social actors 8

emission profiles, alternative 86
emissions, regional composition of 61
employment generation 53
energy 89–93; composition of supplies 93;
 efficiency improvement 90; environmental
 depredations of society 89; intensities 90–2;
 primary requirement by source 150–1;
 production 1, 89; resources, international
 trade in 56; sector, strategies for 85; structure
 of sectoral demand 55; use, intensity of 55
envelope of possibility, *Conventional World*
 scenario assumptions 83
environment 21; criteria for sustainability 39,
 49; damage, long-term 60; degradation 78,
 109; edicts 109; impacts, total 112; link 8;
 objectives 38; pollution in OECD countries
 39; preservation 4; problems, inherited 119;
 problems, contemporary 122; risk 60;
 subsystem 5; sustainability 31, 85
equitable distribution of income 80
equity: impacts 124; international 80, 82, 84;
 national 80, 82, 84; patterns in the scenarios
 26; social, approach to 112; space 26
ethical imperative, new 121
Europe 62
European Union 22, 113
evolution of socio-ecological systems, role of
 different timescales 15

farm products, global requirements for 96
farming systems: irrigated 96; in *Market Forces*
 and *Policy Reform* scenarios, irrigation 170–1;
 knowledge-intensive rather than input-
 intensive 94; rainfed 96; sustainable 94–5
federation of regions 119
feed sources, informal 59
fertility rate 16
fertilization across cultures 118
Fifth International Conference on Adult
 Education 38
final energy-using sectors 91
final fuels 55
Fischer, G. 70
fish and seafood demands 59
fisheries and aquaculture 95
fissionable materials 64
flat corporate structures 22
food: access to 59; and land 93; demand and
 production, changing structure of 56;
 production, increase in 59; requirements 52;
 self-sufficiency 96; shortage, absolute 56
Food and Agriculture Organization of the
 United Nations (FAO) 37, 58
forest: and ecosystem 97; and forestry 72; areas,
 natural 72; loss 97; products 72; use,
 unsustainable 72